The New High-Tech Manager

Six Rules for Success in Changing Times

The Artech House Technology Management and Professional Development Library

Bruce Elbert, *Series Editor*

Applying Total Quality Management to Systems Engineering, Joe Kasser

Engineer's and Manager's Guide to Winning Proposals, Donald V. Helgeson

Evaluation of R&D Processes: Effectiveness Through Measurements, Lynn W. Ellis

Global High-Tech Marketing: An Introduction for Technical Managers and Engineers, Jules E. Kadish

Introduction to Innovation and Technology Transfer, Ian Cooke, Paul Mayes

Managing Engineers and Technical Employees: How to Attract, Motivate, and Retain Excellent People, Douglas M. Soat

Preparing and Delivering Effective Technical Presentations, David L. Adamy

Successful Marketing Startegy for High-Tech Firms, Eric Viardot

Survival in the Software Jungle, Mark Norris

The New High-Tech Manager: Six Rules for Success in Changing Times, Kenneth Durham, Bruce Kennedy

For further information on these and other Artech House titles, contact:

Artech House
685 Canton Street
Norwood, MA 02062
617-769-9750
Fax: 617-769-6334
Telex: 951-659
email: artech@artech-house.com

Artech House
Portland House, Stag Place
London SW1E 5XA England
+44 (0) 171-973-8077
Fax: +44 (0) 171-630-0166
Telex: 951-659
email: artech-uk@artech-house.com

WWW: http://www.artech-house.com

The New High-Tech Manager

Six Rules for Success in Changing Times

Kenneth Durham
Bruce Kennedy

Artech House
Boston • London

Library of Congress Cataloging-in-Publication Data
Durham, Kenneth, 1952–
The new high-tech manager : six rules for success in changing times / Kenneth Durham, Bruce Kennedy.
p. cm.
Includes bibliographical references and index.
ISBN 0-89006-926-3 (alk. paper)
1. Management. I. Kennedy, Bruce, 1950– . II. Title.
HD31.D845 1997
658—dc21 97-3746
CIP

British Library Cataloguing in Publication Data
Durham, Sir Kenneth
The new high-tech manager : six rules for success in changing times
1. Industrial management
I. Title II. Kennedy, Bruce
658

ISBN 0-89006-926-3

Cover design by Darrell Judd
This book is a derivative work. Much of it was originally published under the title *Escaping the Maze.* © 1995, by Quantum Institute. LOC #95-71685; ISBN: 0-9649261-0-5

685 Canton Street
Norwood, MA 02062

International Standard Book Number: 0-89006-926-3
Library of Congress Catalog Card Number: 97-3746

10 9 8 7 6 5 4 3 2 1

Contents

Preface

Sooner or later, every staff person looks at his or her boss and thinks, "What's so special about my boss? I know more about what goes on around here. Given the chance, I could do the job better." Eventually, some staffers push into management to prove themselves. Others are gently pulled into management by circumstance. Regardless of the path taken, they find that leaving a successful staff career to join management can be a rude journey through a jungle of fear, misunderstanding, prejudice, and stereotype. That is why we wrote this book. We want to make the move from staff to management less intimidating and more understandable. And for those who think they have already made the move, we bring some unhappy news: It is not over. Somewhere, someone on your staff is saying, "I could do your job better."

As authors, we cannot help but weave the threads of our own personal backgrounds and experiences into the stories we tell. Because we come from technology-centered industries, our narrative will be particularly appealing to engineers, scientists, and other technically trained staff. Even so, readers with other natural bents will see the relevance and appreciate our principal and repeated lesson—that to become an effective manager no one need learn a new and alien expertise. That is not to say that becoming a competent manager and remaining so is easy work. The changing business environment demands that every manager have a consistent framework from which to interpret shifting requirements and opportunities. This book shows everyone, no matter what his or her professional inclination, how to develop such a framework.

Today, scientists and engineers are moving into leadership positions that for nearly half a century had been reserved for other professions. Some watching this great migration welcome it as a return to the logic, clarity, and discipline common to technically trained people. For others, it represents an unwelcome return to the conceit, intolerance, and boorishness common to technically trained people. Regardless of how you feel about the situation, the fact remains that technologists will be increasingly called on to guide the businesses their inventions, discoveries, and applications create. This change in leadership perspective will be felt up and down the watchtower.

It is natural for business leadership to be dominated by one profession or another. It has been so ever since the first banker financed the first sea captain using other people's money. The movers and shakers in any company are chosen for how well their talents match the opportunities in the marketplace and how well their pedigrees compare to those who traditionally have held power. Few people were surprised that in the 1990s the heads of newly incorporated, privately held Russian companies were usually ex-communist apparatchiks or organized crime confidants. America's business history also shows definite leadership-preference patterns, starting with owner-capitalists who built vast family fortunes and ending with the rise of the professional manager charged with protecting the interests of absent owners.

The predominate background of professional managers has changed markedly over the last few decades. Telling the story of how technologists lost their influence in American business and are now reemerging is interesting but of little practical value to those who are experiencing the shift first-hand. At the same time, we want to avoid the common mistakes of books intended to educate management. Either they are too narrow and theoretical in focus, like the many books aimed at marketing and accounting functionaires, and are more appropriate for the practicing specialist. Or they are written by newly retired and widely admired CEOs; those books tell war stories and warn about problems ahead but offer no real-time action plan. Or they pump up the newest management fad, such as the hundreds of books that push the latest quick-fix business fad. The world is full of self-help management books talking about management, but few talk about the *job* of a manager. We want to give you a practical understanding of the tactics of the campaign ahead and do so with a hint of irreverence fortified with a heavy dose of contrarian thinking. Also, we believe that science and engineering educators will appreciate a plainly written text suitable for a management survey course designed for students whose career path might pass through general management's giant gravitational field.

We take it for granted that technologists want a consistent, coherent management skill set that complements rather than repudiates their previous education and experience. To that end, we begin with the self-evident truth that engineers and scientists look at the world differently than those who grew up in other professions. That difference is a consequence of disposition, environment, and the personal compromises and rigor necessary to master a technical subject. That difference becomes obvious when technologists begin to make the transition from their traditional staff jobs to management responsibilities. While we axiomatically accept that technologists are emerging as the chosen occupants of professional management, it is our premise that much of what makes an engineer an engineer and a scientist a scientist is also an appropriate foundation for managing an enterprise. That proposition is gaining favor because of the media's infatuation with our increasingly technological world, but the intellectual ingredients supporting that claim have always been around. Only now, it is an easier sell.

The source of most scientists' and engineers' difficulty in making the transition to management is not their preference for solving problems with tools, models, experiments, and trial and error. It is that they abandon their problem-solving preferences for the incantations of professions that preceded them, such as marketing, law, or finance. That self-imposed abdication is supported by widely recognized professional stereotyping. Everyone knows that technologists do not know the first thing about human nature, human organization, or the human condition. They cannot hold a social conversation, they do not know how to motivate people, they seek protection behind their technology, they are clueless about politics—all supposedly important skills for managers. Perversely, many engineers and scientists actually enjoy being seen as so. They make jokes about their own inadequacies and entertain themselves with witty stories about the less technically enlightened. It is a tribal thing. It helps to have a sense of humor about those conventions in the face of limited careers and diminished roles. The outsiders, that is to say, the technically challenged, have their own mechanisms for reinforcing the stereotypes. They admit to their engineer and scientist collaborators that, yes, they once wanted to be engineers, too, but hey, who wants to live life as a nerd? To the techie tribe elders, it is just sour grapes. The outsiders just could not hack it. If they could, everyone would be an engineer or scientist, right?

It is obvious that this defense mechanism serves both sides. But can we really say that scientists and engineers are less equipped for social intercourse than their brethren in other professions? One need only look at the vast amounts of energy technologists spend building alliances within their own departments to realize that there is more here than just a bunch of pocket protectors. The fact of the matter is that a person's inherent management style is the consequence of his or her total identity, not just one part. So there is nothing keeping engineers from being as effective as those who come to the job of management with a different orientation. What is different is that each profession approaches basic problem solving differently. When engineers or scientists first encounter the world of management, they detect an irrational but dominate management paradigm, presented in a language they do not understand, full of internal inconsistencies, and definitely outside the range of their previous analytic tools.

We believe, and will shortly prove, that it is possible to build a set of fundamental business skills from a perspective that technologists appreciate by making it correlate with their traditional methods of solving problems. Yet this approach is friendly to anyone interested in a more structured approach to the job of managing regardless of background. That led us to coin the word "manageering," a word that emphasizes the roots of engineering (and scientific) disciplines that anchor management science. Management science certainly has other roots, so some readers will be unhappy with our particular selection of fundamentals. But manageering works, not because it is irrefutable but because it fits the new, emerging class of managers.

Chapter 1

The Fundamentals

A strange, almost perverted thing happened on the way to our modern, high-tech business world. Many technologists found themselves enthroned as Knights of the Lab, while the Suits laid claim to King of the Company. They all used to be very, very happy with that arrangement. The lines of demarcation were clear, Knights and Suits each had to pay an appropriate level of attention to the other, and in times of trouble they thought they had the luxury of pointing fingers at each other. If you look carefully, however, you will find undeniable evidence that this once-cozy relationship is changing. But first, let us review how these familiar features of the management landscape came to be.

Over time, the keepers of technology convinced themselves that technology had become so complicated that only they could understand it. In turn, the Suits convinced themselves that modern business had become so complicated that only they could understand it. Each camp was pleased with that division of labor. The scientists and engineers were left to pursue their crusade of taming nature, to marvel among themselves at their superior intelligence, and to read about themselves in their peer journals. The Suits were more than happy to be left to make the deals, to marvel amongst themselves at their latest winning corporate strategy, and to read about themselves in the business press.

In the mid 1980s, a tacit recognition that not all was well with that division of leadership began to coalesce. Lo and behold, some of the most successful American corporations—Microsoft, Intel, Compaq, Exxon—were headed by scientists and engineers, not marketers, financiers, or lawyers. The elevation of the director of technology to the vice-president's office also hinted that we were going back to our industrial past, a time when founding technologists both formed and ran their own companies. This back-to-the-future phenomenon was happening in other parts of the business landscape, too. For instance, many people rediscovered the role quality assurance plays in business survival and success, thus spawning legions of consultants ready to reeducate us in the fundamentals of that nearly forgotten profession. The emergence of activity-based accounting from a long, 50-year slide into oblivion was another indication of the revival of fundamental business practices.

While the general gravitation back to business fundamentals is a fascinating story, the move to entrust technologists with the general management function has more profound implications for American industry than perhaps any other recent business phenomenon. Why? Since the 1940s, America has gone through several management eras, and each has left an indelible mark on our economic history. Each era can be defined by the predominate background of the CEOs of the time, and it is that shared point of view that shapes contemporary business thought and sets its direction. Roughly speaking, first came the mass marketers, people trained in gauging contemporary tastes and packaging them. We suffered through the lawyer period, when boards realized that vast amounts of money could be made or lost in the courtroom. Next, we have seen management ranks taken over by financiers who figured out ways to drain cash from market goodwill. Those shifts reflect both a reaction to opportunity and a faddish response to leadership selection. Nonetheless, for those very same reasons, American business is now entering a new era marked by the ascendancy of the technologist as manager. However, on a personal level, that shift is more easily said than done.

1.1 INTRODUCING THE TECHNOLOGIST-MANAGER

When you meet new people, they invariably ask you what you do for a living. You and your new friends use euphemisms to describe what you do, but regardless of the words, most people will form a fairly straightforward and graphic picture of you at work. Suppose you are a surgeon and tell them so. They immediately form a picture in their minds: They see you cutting people open, rearranging what you find, sewing them up, and seeing them through recovery. Say you are a pilot, and people see you using your skills to guide a plane full of people or cargo through a complex air-traffic control system, getting from point A to point B. They see the plane, the airport, the uniform. Suppose you are a "manager." What picture would your new friends paint? What skills would they think you have? What does your finished work look like? If you explain yourself by saying, "I'm good with people...I understand numbers...I solve problems...," you get a polite, but nonplussed, "Oh, I see."

To make your life easier, instead of saying you are a manager, you might tell them what you used to be, as if you still did that job. You could say you are an engineer, an accountant, a salesperson, a lawyer, or a programmer. You know that if you chose a familiar title, people will be able to form a fairly coherent mental picture of your work without much explanation, even if the picture is wrong. For example, if you own your own company, you feel safe to say that you own a company. People understand ownership: They see someone in an office reading mail, taking phone calls, relaxing over early lunches, and taking off in the afternoon. The conversation then turns to what the *company* does, not to what *you*, the owner, do. The conversation gets over the first hurdle nicely. If you cannot claim to be the president or the owner but nonetheless have similar management responsibilities, it is likely you will mention your former profession.

There is nothing wrong with your little white lie, reclaiming your previous profession just so you can keep party conversation going. However, some managers explain their work in terms of their technical expertise because they never have come to terms with being a manager. At the end of most days, they feel they have accomplished something. But ask them to explain how they do what they do and what exactly it is they do, and they will be able to relate only to the technical side of their job, or about 20 percent of it. Eventually, that 20 percent claims 100 percent of their attention because the technical side is the only part of their work they know. When faced with a critical management task outside their technical realm, they find themselves lacking the other necessary skills. They come to lean on ready-made, fashionable business fads, joining the ranks of camp followers rather than leaders. It does not have to be that way.

1.2 WHAT IS A MANAGER'S REAL JOB?

Before we can list a manager's fundamental skills, we have to define the job. That's fairly simple: Managers run the business. How do they do that? They ensure that the operating systems defining their department or business are designed, installed, and maintained. After all, what is a business if not a collection of agreements between people concerning how to organize their individual work such that they can sell a product or a service for a profit? Those agreements are collected into various systems—the order entry system, the engineering and design system, the financial reporting system, the production control system. *System* comes from the ancient Greek *syn* and *histanai,* meaning "to cause to stand." Translated into the business world, the system is "that upon which the business stands." Without a systematic way of doing work, a business cannot stand. Someone has to be responsible for making sure the system works, and that someone is the general manager. Managers are no strangers to systems and refer to them every day. They are not mysterious conventions invented by people long forgotten. They are alive.

This leads us to debunk our first myth. It is popular today to say that the system belongs to everyone, not to management. However, individuals who work *in* a system are at a disadvantage when it comes to working *on* the system. Working on the system requires straddling the line between work as it is done today and work as you want it done in the future. That maneuver takes a suite of complex skills that are accumulated over time, reinforced by constant application, and supported by appropriate rewards. If you give people who have been working in the system the responsibility and authority to work on the system but deny them adequate time to learn the new job, deny them the opportunity to try different approaches to changing the system, and deny them a new role in the new system (by indicating that they will return to their "old" jobs when the project is completed), you are finished.

The job of management is neither a part-time assignment nor a temporary occupation. The system on which your business is built needs constant care and professional attention. To understand those needs and to be able to provide for them take the thoughtful training and insight provided by the pursuit of a set of fundamental skills.

1.3 WHAT IS FUNDAMENTAL ABOUT FUNDAMENTALS?

First we want to be sure we have a common but precise definition of a "fundamental skill." Let's look at *skill* first. The dictionary defines *skill* as a "developed aptitude or ability to do something competently." Now, the definition of fundamental. If it is fundamental, such a skill determines "essential structure or function." So by definition, if we develop a list of fundamental skills for a manager, we have defined the essence of what it means to be a competent manager.

Moreover, we must recognize that any skill takes practice to develop and requires repetition to maintain. For instance, a pilot must fly to keep sharp. A surgeon must perform surgery to remain proficient. Not many of us would board a plane knowing that the pilot had not flown for six months. Few of us would be happy being cut open by a surgeon who had not performed any surgery in the past year. All professions demand a significant level of attention to some set of fundamental skills. Nevertheless, it is fashionable to empower line personnel to work on the system that regulates their work, even though they lack a complete set of fundamental skills to do so. Why? Why are newly minted managers expected to bring an undefined skillset to a poorly defined job? And why do managers expect themselves to remain competent if they neglect their fundamental skills? It is irrational.

1.4 IS THIS JUST ANOTHER LIST?

We are not claiming any new insights with our list of six fundamental management skills. Fashionable management fads invariably claim that they have discovered new truths. After you slog through such a program, the truth you find is neither novel nor profound. At its core is one, and only one, of the six ideas. While the fundamentals we present in this book are not new (if they were, they would not be fundamental), our collection is displayed and explained in a coherent framework. Like any collection, it has to be seen as a whole, not as pile of individual pieces you can pick through. A common demand of art collectors, for instance, is that their collection be kept together when they bequeath it to a museum. The same applies to this collection of fundamentals. You cannot divide it. No one item can be taken from it without changing the appearance and meaning of the remaining items. One of the reasons management fads fail is that they take one of these fundamentals, adorn it with all sorts of jargon and superfluous activities, and claim that if you apply it in a program of good hygiene and regular professional care your problems will be solved. They rarely are.

Even though we have not yet explained what is behind our six fundamentals, you should be suspicious of anything that claims to be fundamental, much less something that claims to be fundamental to a discipline that many people think is either hardly a profession at all or one that can be learned on the job. Being a good manager is a difficult job that requires fundamental skills, but how will you *know* after looking at our list that its elements are indeed fundamental? A practical, real-world test of any fundamental idea is finding out what happens if we ignore it.

For example, take the ancient philosophical argument about whether reality is real or just a construct. As children, many of us wondered whether the world was put here for our individual entertainment or if it was something bigger. How could we tell if reality was real or if the world as we perceived it was a fabrication of our minds? That problem is a no-brainer for philosophers who call themselves realists. It is pretty much a no-brainer for the rest of us, too. If you ignore any of the fundamental elements of reality you will get hurt. Suppose you want to test the substance of reality by stepping off a cliff. If you take one step over the edge, the practical result for the rest of us is that you are dead. For those of us who remain behind, reality is a fundamental truth. It is the same for any list of fundamentals, whether it is the one we present here or it is someone else's. You face metaphorical death if you ignore the fundamentals.

The current enchantment with the fad of self-directed, empowered teams is a good test of fundamentalism. The premise is that the best method for solving problems is gathering together those people with intimate understanding of the situation, training them in problem-solving techniques, and letting them loose with little direction. They are allowed to define their own problems and generate their own solutions. Tales of success and triumph of this problem-solving method saturate the business press. Whole companies have been saved and industries revitalized! But is knowing how to assemble self-directed teams a fundamental skill? As you read on, it will become obvious that while teams-building might be a skill preferred by some, it is not a *fundamental* skill required by all. It does not represent an essential structure or function of general management; it represents a style. If you remove this skill from the face of the earth, businesses will not perish. But if you ignore any of the tried and true fundamentals we present, you are in real trouble.

1.5 THE SIX FUNDAMENTALS

We chose the term "manageering" for our list of fundamentals because, at its most basic level, it represents the conjoining of management and engineering. We will not, however, be repeating the word "manageering" like a mantra throughout the book. It is more subtle than that. We want you to be conscious of something you already feel subconsciously, that engineering and science can form a solid foundation for assembling fundamental business skills. More important, the implication is that since managers are concerned with the health of the business as a system, a manager with science or engineering bent, who therefore is already knowledgeable in the language of systems, has a jump on the competition.

What follows is a simple sketch of each of the six fundamentals. We will give you enough to get you started but not enough to steer you to the first landmark shown on your map. This is your shakedown cruise, so the initial explanations may seem incomplete. They are not intended to last the entire trip. The chapters that follow will provide the sustenance you need to be comfortable during your travels through the thick jungle.

1.5.1 Profit Awareness

The primary focus of general management is profit. That statement is so fundamental to business it seems trite and out of place in a list of fundamental management skills. Of all the fundamentals we discussed while writing this book, the decision to include profit generated more disputes because it is so fundamental as to be axiomatic. But think about your own experience. When was the last time someone mentioned profit when discussing competing solutions to a problem? Sure, you hear a lot of words about improving design, reducing redundancy, enhancing quality, or increasing market share, but when was the last time you actually heard the word *profit*? You probably cannot remember; if you do, either it was a rare event or you work in a rare environment.

This absence of the word *profit* is not a trivial matter. It is not merely an oversight, either. It is as if a mariner has lost map and compass. The ship might be guided to landfall by way of the sun and stars, but arriving at the chosen destination would be blind luck. In our case, if such a fundamental concept as profit is not spoken aloud, often and vigorously, its power is soon lost. As we said, because the profit motive is the bedrock of business, it almost did not make our list of fundamentals. But its absence from contemporary business vocabulary demands that it be on the list and have a special place: the cardinal skill. That we had to include a skill akin to breathing is a sad testimony to the extent that management-by-fad has replaced management-by-fundamental.

1.5.2 Technical Literacy

It is fashionable these days to insist that managers do not need to know the technical foundations of the company, division, or department they lead. That notion is one of the biggest intellectual frauds of the past 20 years. That people readily accepted such a rationalization for technical incompetence in the management function shows that, like any fad, that specious idea had a ready market.

As we explain later, your level of technical competency need not reach that of your most proficient technician, engineer, or scientist, even if you have the capacity to reach that proficiency. In fact, it defeats the idea of division of labor if you expect yourself to stay abreast of any but the major technical developments that affect your business. However, general management does require something called technical literacy, a kind of knowledge readily available to anyone with the basic intelligence required of successful managers.

1.5.3 Change Dynamics

Controlling the direction, pace, and impact of change is such a fundamental management skill that it, too, is in danger of becoming a fad pushed by experts with ready-made programs. Remember, the core concept of a business fad is not necessarily bad. It is that the fad overwhelms the fundamental ingredient to such an extent that the original concept is unrecognizable.

Change dynamics include issues of "culture" (another overused, misused, misconstrued, and generally abused concept), human and organizational behavior, and change measurement. Without a coherent framework that explains how people and organizations change, you will not be able to manage the process consistently and effectively. While researchers can spend their lives plowing that fertile field, you will be interested in learning just a few basics of how companies change.

1.5.4 Policy and Procedure

It is disappointing how many otherwise competent managers believe that establishing policy and procedures is a necessary evil. The arguments against writing policy and documenting practice are easily recounted. You probably have your own list collected over the years, including the following: "We do things differently every time." "We are a unique company in a unique industry." "Our technology is so different that policies and procedures do not work." "We do not have time to write everything down." "We do not want to be burdened by paper." "If you start down that road, you will need a piece of paper to turn off the lights."

People who shudder at the thought of reducing policy and procedure to paper nevertheless expect their paychecks to be issued on time and error free. They expect consistent service from the grocery store. They demand that government services be cheaper and more efficient. You can eventually get them to admit that if they did not get their checks on time, their groceries bagged properly, or their mail delivered every day, the solution would include establishing a few procedures and keeping some records. Beneath the denials, we know that without paper and pencil (or their replacement, the computer) a business could not run.

So, let us agree: The debate is not about whether records, procedures, and policy are necessary but about how much is necessary. Taking the position that reducing policy and procedures to paper is de facto a bad idea is a nonstarter. To overcome that fashionable, knee-jerk attitude, you need to learn how to determine what must be documented and what does not. As you put these ideas to work, you will begin to appreciate that written policy and procedure can make you money.

1.5.5 Decision Theory

At first blush, it is strange to think that you have to learn to think. But of the many skills we must learn in life, our teachers do the poorest job teaching us to think. Organizational behaviorists and clinical psychologists have a name for the process of thinking about alternatives and choosing one over the other: decision theory. Many of their ideas about how good people make bad decisions are perhaps the most useful tools you will ever learn as a manager. This book explains some of the more prevalent traps in decision theory. You will be astounded that by merely changing the way you ask questions you will get better information and have higher confidence in your decisions.

1.5.6 Duality

Strictly speaking, duality is "the state of being of two irreducible modes." The two modes are (1) meeting the needs of the group and (2) satisfying the needs of individuals, including the managers themselves. We know intuitively that any system with two or more irreducible modes has a tendency to become unstable. We never know which mode will manifest itself. In some cases, one mode cancels the other, at other times each enhances the other. Sometimes the tension can tear the whole system apart. The term "duality" expresses the essence of that balance, not only because every manager has to operate in two modes but because the conflicting modes (or roles) can literally destroy a manager's effectiveness.

Walking the tightrope between the needs of the group and the needs of the individual takes someone who understands the dual role of leader and comrade. Thankfully, mastering duality is a learned skill.

1.6 GET READY FOR YOUR TREK

For a manager, learning the fundamental skills is equivalent to a survivalist learning cross-country trekking skills. Maps are helpful, sometimes crucial, but not absolutely essential. What is essential is a good grounding in fundamental skills. They alone ensure that you will avoid running in circles, exhausting yourself and eventually getting lost in a seemingly unfamiliar world.

Chapter 2

Profit

Among any collection of fundamental skills there is always one cardinal skill. For instance, take baseball. All players (except pitchers) have to have a "quick bat." Regardless of position, if players are going to hit in the big leagues, they have to learn to get their bats around very, very quickly. A big league pitch comes at them so fast that the time between the decision to swing and the swing itself is nearly instantaneous. Sure, ball players need other skills too, but without that cardinal skill, they will not hold their jobs. So what is the fundamental skill of really, really good general managers without which they would not be able to hold their jobs? The ability to keep focused on profit.

2.1 THE TRIUMPH OF POLITICAL CORRECTNESS OVER MAKING MONEY

The goal of your professional life's work is generating profit, period. That may seem like a straightforward ideal, yet for some, profit is a dirty word. That feeling, generated by a small, elitist faction in our society, can subdue those charged by society with making money. Sure, not all people pursue profit as their life's work. There are plenty of other noble goals to work toward. But the cold, hard fact is that in this world most of us work for "the spread". Once you are profitable, and consistently so, you can do all the things wealth allows, including rejecting it. Sure, most of us would like to be wealthy enough to reject wealth. But until then, most of us are one month away from the virtual jungle.

Because of that emotional overhang, people are always trying to find ways to avoid saying the P-word, but there just is not a very good substitute for it. For instance, in 1995 a big flap developed during the planning for anniversary celebrations of the victory of the Allied forces over Japan during World War II. President Clinton was purported to be reluctant to refer to that historical day as "VJ Day," believing that to do so would be discourteous toward the Japanese, now one of our closest diplomatic friends. During a press briefing surrounding the VJ Day flap, a White House aide was quoted as saying defensively, "VJ Day, VJ Day, VJ Day, VJ Day, VJ Day! There, I said it five times in a row."

Our point is this: Is the word "profit" your "VJ Day"? Do you have trouble saying the word in polite conversation? Many people are not comfortable using it. Given a choice, they would rather defend the righteousness of a male-dominated society than defend the virtues of a profit-driven motive. Unfortunately, that politically correct discrimination has a deadly side effect. People begin to avoid using the P-word when and where they absolutely must—in the impolite office arena. Not only should you use the word "profit," you should use it until the scab of political correctness falls off.

2.2 PROFIT IS OUT OF SIGHT, IT IS OUT OF MIND

Recently we met with a group of owners and managers of an emerging technology company. We were discussing the role of policy and procedures. When we asked them to list the reasons they were spending their time at this meeting, only one of the nine mentioned profit, and then only obliquely. The sad thing is that such a response is typical. Few businesspeople keep a portrait of profit anywhere in their foreground. What they do is use lots of surrogates for profit. They will say their new initiative will improve customer service, reduce overhead, increase sales, improve quality, facilitate teamwork, or reduce cycle time. Never does it occur to them to say the new program will increase profit.

Why is it so important for managers to be more comfortable with the P-word? Because a surrogate is never as good as the original—*never*. Why? Because when the surrogate goal replaces the real goal, you have a subsequent loss of focus. Look what happens when a fad tag like ISO 9000 or reengineering replaces the goal. Severed from the real goal of making a profit, the management fads take on a life of their own, and profit suffers.

2.3 PROFIT IS NOT A FOUR-LETTER WORD

All business failures are ultimately about losing focus on profit; all business successes are ultimately about sustaining a focus on profit. All other explanations are merely expansions on that theme. So if you find that surrogate concepts have taken the place of profit in your business vocabulary, you have a serious infection. You have to get over it, but how?

Try using the word *profit* in every answer to every business question for a solid week or two or three. You will be surprised how easy it is to start every answer with "Let's discuss how each alternative affects profits." Eventually, this little exercise will become second nature, and you will learn to demand that people frame their alternatives in terms of improving profit right from the start. While it certainly will feel good to become focused on why we come to work in the morning, this transformation also helps technicians communicate across knowledge boundaries in a common and richer language. Unknowingly, they will quit treating *profit* as a four-letter word. Eventually, they will have no problem using it in polite conversation.

2.4 AN EXAMPLE

Let us look at a real-life example of how focusing on profit clarified a problem. A computer system administrator claimed that the current stand-alone computer work stations in the design department needed to be replaced with a network. She reasoned that a networked system would improve efficiency because design engineers no longer would have to trade floppies, thereby eliminating the confusion and cost surrounding the question of which of many revisions was the latest. All that was fine and good, but several questions begged to be answered.

Later, we will discuss the importance of demanding alternative solutions, but for now we want to show how important it is to ask questions about profit. We asked, "How much money will you make from the investment? How many months will it take to pay it back? What is the primary source of the profit from this cost?" If your staff is typical, they usually present their ideas in terms other than profit. They will be at a loss to answer questions about profit. But if, after struggling with the answer, they cannot develop a rational set of numbers showing how the investment in new technology will make money, then you should not approve it. Why? Because you cannot generate numbers unless you can describe exactly how the investment will be used. If you cannot describe how the investment will be used, you are not ready to take advantage of it.

In our example, the system operator's first stab was to quantify the unquantifiable. She made some meager assumptions about how often files were misplaced, how often the wrong file was updated, and how long it took to swap diskettes between users. Most of the evidence was anecdotal, but at least it was a start. We then asked her to set up a rational method for gathering actual data about those activities. By then she was well on her way to developing a robust, profit-centered picture of "improving efficiency."

In the end, the investment in the network was justified solely on standardizing print size and reducing paper costs. Nuts and bolts, dollars and cents. The fact that other problems might be solved did not enter into the equation. Efficiency improvements and better communication are hard to quantify and easy to dispute, and unfortunately they rarely materialize. The real value of this particular investment was not the network per se but the new printing technology that required a network environment to support it.

2.5 THIS HAS A FAMILIAR RING

Technologists will recognize something familiar in the cadence of this solution: Generate a hypothesis, construct an experiment to prove the hypothesis, run the experiment, gather and interpret the data, and either start the process over again if the data prove you wrong or go on to the next problem. This last phase of problem solving separates science from business. Scientists and engineers like to think that problems will be solved eventually, if not by the current generation, then by those that follow. That is part of the myth of the march of science and technology. In business, the rules are different. The overriding concern is that there are more problems than money or time available to solve them, and you can spend too much on the wrong problem. So managers are responsible for deciding how much time, effort, and money can be spent on the problem at hand in relation to the other problems that need attention. All companies have problems they need to work on but do not have the necessary resources. Subsequent chapters address ways to solve this real business dilemma.

2.6 FALLING IN LOVE WITH TECHNOLOGY

This is a good place to introduce a theme that we will come back to later from another direction. It is our experience that when investments in new technologies, especially computer technologies, do not return money to the bottom line, they do so for the one principal reason. No one took the time to apply the same intellectual discipline to those decisions as they would to buying a new machine for the production line. Those with a technological bent do not seem any better than anyone else at avoiding that problem. The power of the first manageering fundamental—a focus on profit—derives from its being applied consistently over a long period of time to all decisions.

All decisions—whether about hiring, firing, expanding a product line, dropping a line, setting up a booth at a convention, increasing employee benefits, decreasing benefits, outsourcing, starting a new management program, hiring your brother-in-law, resurfacing the parking lot, holding a Christmas party—need to be framed in terms of increasing your profits. Can you justify it in hard numbers? If not, then do not do it.

As an exercise, think about how you can justify the value of a company picnic, using dollars and cents. Would a picnic paid by the company make money for the shareholders? Your answer had better be "yes," and your calculations should show an exact amount.

2.7 NUMBERS, SCHMUMBERS, I CAN GET THEM TO SAY ANYTHING I WANT

One last point about profit measured in dollars and cents: On an incremental basis, seldom is the actual number important by itself. The real meat and potatoes is what you discover about the problem itself during your analysis, especially what intelligent things you can say about the problem's sensitivity to a change in inputs. We chose the company picnic expressly because through it you can discover the paradoxical nature of so many management problems. Behind the issue of a company picnic are several questions: How do we measure company morale against company-paid benefits? How do we determine the appropriate role of management in employees' lives? How do we justify the use of company assets for what, for many, appears to be personal use? And how do we compare the profitability of a company picnic to the profitability of other perks, such as supplying chauffeurs to upper management? The answers to those questions must be put in dollars and cents, because to make economic choices about the company's assets requires answers in dollars and cents. In the end, it is all about profit, man.

As for sensitivity to inputs, scientists and engineers use the concept of stability to express how a system will react to a spike in one of its inputs. The same idea applies to project-economic analysis. Every business problem starts with the inputs of the cash, time, and effort required to implement a solution. What happens if the cash requirement doubles, the time to completion expands to fill the time available, or management oversight requirements suddenly go from slight to out of sight? How will any of these possibilities affect all the other projects that need your attention and a piece of your assets? A profit model will help answer those questions, and the answers will give you a feeling of the risk of the undertaking.

2.8 SITUATIONAL FUNDAMENTALS ARE NO FUNDAMENTALS AT ALL

Figuring out how much money you will make on a company picnic is, of course, a radical (some would say, bogus) exercise. But if focusing on profit is a cardinal skill, we cannot shy away from the sticky problems that reside at the margins of business and life. Besides, you have to show yourself that fixating on profit really is a fundamental skill and that you can do it. But remember what we said about the list. It is a collection, and no one element stands alone. That does not mean that one element can be ignored since others will fill the gap, nor that one is so powerful in certain circumstances that it holds sway over the others. A collection of fundamentals cannot be affected by "special situations." It means that when you work a problem you have to apply all the skills to solve it creatively, completely, and absolutely. If, after thinking about quantifying the profit of a company picnic, you come up empty-handed, do not worry. The other fundamentals will help you out.

Chapter 3

Technical Literacy

The second fundamental skill of management is technical literacy. Managers must be able to cut to the chase on technical issues, which means they have to retain a minimum grasp and cultivate an informed skepticism in a wide range of applied disciplines. Why is this skill, among the many that good general managers must possess, a fundamental one? Because managers are a company's "techno-linguists," and the general manager is the company's chief techno-linguist. To be able to get things done as a manager, you have to communicate ideas to a wide range of technicians using language you both understand. But it is your job, not theirs, to be multilingual because you are the one who has to mediate between technicians who have competing sets of ideas and speak different technical languages.

This technical faculty is different from that of technicians. For example, to live and work in a foreign country, you have to speak the language. You will not speak as well as a native, nor are you expected to. So relax if you are technically ignorant in some facets of the work around you. There are ways to pick it up, and we will introduce them to you. But before we go on to outline how that can be done, we need to discuss what technical literacy is and what it is not.

3.1 MANAGERS NEED TO KNOW BUSINESS, NOT TECHNOLOGY

Our idea—that managers' jobs depend on their ability to translate technical expectations across functional boundaries—flies in the face of a 20-year movement to remove the technical component from general management. The idea that general managers need not know the technical side of business nor understand the core technology is a popular one. It is said that, after all, managers just pull the strings; thus, the only fundamental skills they need are people-manipulation skills. Right—and a music conductor does not need to know how to play an instrument or understand acoustics to run a successful orchestra. That mindless drivel was taken to the mountaintop and clothed in the latest business fashion by a popular 1990s management book claiming to illustrate the common traits of really, really successful people. A whole consulting industry was built around the implication that by copying those traits you, too, could become fantastically successful. Some Fortune 500 companies inculcated their people with this new business religion.

It is interesting that when this New Wave book first came out it was aimed primarily at the self-help market and could be found in the personal improvement section of bookstores, next to such self-help classics as *I'm Okay, You're Screwed Up* and *If Life Gives You a Lemon, Demand a Refund.* But there is a basic problem with trying to copy character traits—they cannot be copied. By definition, *character* is something you have, not something you build through exercise. Everyone has an opinion about character formation, but no one knows for sure how we develop our own distinctive characters. Some observers speculate that the process is pretty much complete by the age of five or six; others assign its completion later in life. But most people agree that character is certainly set by the time we start to work for a living. So you can no more become a successful person by trying to copy a set of character traits than you can become a professional athlete by copying an exercise regimen.

It is a myth that management style determines management success. The fact is that really, really successful businesspeople are the most diverse group on the planet. The fact is that successful persons have their own individual styles. Taskmaster or integrator, you will find plenty of people who like your style and want to help you reach your goal, but only if you know who you are, what you want, and how to communicate it. Regardless of your style or your character, in this world if you are technically illiterate, you will not be able to talk to many of the people who will determine your success. And if you cannot talk to them, you cannot get any real work done.

3.2 A PLANE FULL OF EXPERTS FROM KATMANDU

Most of us come to general management via success in another function, such as sales, finance, engineering, operations, administration, or law. Up to that point, we have spent a considerable amount of our lives building technical expertise, which, in our new job as general manager, will only contribute, at most, 20 percent to our continuing success. Unfortunately, even though those skills have little to do with success in our new job, we find ourselves falling back on our old profession. After all, that is where we feel comfortable. Most general managers with engineering backgrounds have their best relationship with the engineering department. It is the same for the ex-salesperson, ex-lawyer, and ex–chief financial officer. They relate best to their old colleagues.

Managers are only human. They try to ignore functions they know little about. Fear is a natural result of ignorance, and overcoming ignorance is hard, painful work. They believe that since they are managers they are paid to have thousand-dollar answers. They figure if they do not have quick, glib answers, then they are frauds. That makes some general managers avoid functional experts, since the experts sound like they just got off a plane from Katmandu.

As general managers, it is easy to be convinced that the more arcane functions can be left to the "experts." They think, we will just rely on our good judgment about people to keep us out of trouble. Besides, there are a lot of apologists who rationalize that behavior. Everyone says technology is so complicated these days that no one person can possibly understand it all. Anyway, is not empowerment about letting people work out their own problems? It really is not our job to babysit these people we have hired. They are adults. We pay good money for them to know their jobs. But wait...look at the reality of business and people. Empowerment as applied in the real world is mostly superficial. For the most part, the only business functions that are empowered are those about which the manager knows little. The functions in which a manager is versatile are kept on a short leash. The general manager who took empowerment to its logical conclusion would reduce the job to inane boosterism. Boy, now that is a job I want!

The hot, harsh heat of reality makes these excuses and inconsistencies wither away. While it is true that technology can at times be unfathomable, its effect on your organization is real and easily comprehended. The corporate battlefield is littered with companies that blew apart at the seams because their managers were functioning technical illiterates. Looking back at the destruction, the managers' abandonment of the battlefield for the rear bunker was obvious to the people who were left leaderless. At the very least, companies in this situation tread water while trying out different general managers, all who look like they have the "right" character but who have not a clue to the company's technology and therefore what is important to its business.

3.3 EVERYBODY HAS GONE HIGH-TECH

When we talk about technology, what are we really talking about? In a world of high-tech hyperbole, it is easy to think that you are not into high technology unless you are into computers or communications or aerospace. That simply is not true. No matter what industry you look at and no matter what function you scrutinize, everybody's job has gone high-tech. From the rancher who tests the nutrients in the grass to determine optimum herd size to the accountant who devises new methods for capturing cost- and value-added activities, they all have gone high-tech. High tech is just another name for progress.

The reference point is not absolute but relative to the previous technical generation. For instance, around the turn of the last century, changing the transportation system from a horse- and steam-powered one to internal combustion was a great and genuine technical challenge. If it were happening today, we would call it "going high-tech." But notice, it is not the substance of the change that makes it high-tech. It is our reference point, and it is all in the eye of the beholder.

Even if you are a technically literate person, when you move into management you have got to develop a completely different technical literacy. You will quickly figure out that you will not have time to stay abreast of all the changes going on in your past profession, so you naturally adjust your literacy level to one of appreciation rather than comprehension. As your problem set gradually expands to include issues from other disciplines, you have to find time and energy to accumulate technical literacy in totally new areas. The level of understanding new problems will rise to the level that your mastery of your old technical profession must fall. Since everyone coming into management comes trained in some sort of technology, everyone has to work hard to reach and maintain technical literacy in all the technical functions with which they come in contact. The only advantage that an engineer or a scientist has over others facing a move into management, if any, may be their unique point of view.

3.4 SOME EXAMPLES

Let us visit two colorful examples of the consequences of technical illiteracy in management. The first is a classic example of managers making decisions without a firm grasp of the technology involved. The second is an example of what happens when someone thinks the answer to every problem is a technological fix.

In 1995, the U.S. Federal Aviation Administration (FAA) admitted that it had spent at least half a billion dollars and 15 years building a new air traffic control system that had an uncertain completion date and that when, if ever, completed would be obsolete. Pulling the plug on the project, the FAA replaced it with new, supposedly more modest $800 million dollar undertaking projected to take an additional 10 years to complete. This story is not one of technology overtaking itself but of FAA management falling in love with technology they did not try to understand; of users making and re-making wish lists full of fabulous features and bodacious capabilities; of vendors encouraging goals, objectives, and timetables impossible to meet; and of technical staffs overwhelmed by the complete and seemingly capricious reversals of project objectives. Simply stated, there was a complete and total breakdown in technical awareness along the chain of command. Obviously, the managing directors of the FAA could not be expected to know the intricacies of the new system, but had they been well versed in project-management technologies and decision analysis they would have avoided a boondoggle of such proportions that only an economy the size of the U.S. economy could endure it.

On a lighter and much less expensive note is the story of Dolly Parton's first attempt at a television situation comedy. In the mid 1990s, Ms. Parton, a successful American singer and entertainer, had to cancel her anticipated half-hour comedy show after producing six episodes. The show was such a creative mess that the six installments were never broadcast. One of the bigger aggravations was that Dolly taped her show in Florida (to be closer to her home in Tennessee) while production executives were 3,000 miles away, in Los Angeles. "Not to worry," said those who should have known better. "We have technology that will solve that problem." The California people could watch rehearsals via satellite in real time and offer their comments and changes via conference calls and faxes. The problem was that the creative process is more than watching a tube and sending ideas down a wire. You have to interact with people. In this case, the lure of slick telecommunications technology overwhelmed the basic question of how the nature of the process might be influenced by the new technology.

The second example illustrates that technical literacy is not the same as technical facility, that is, familiarity with the nuts and bolts of the technology (although literacy does not exclude facility). Literacy implies competence, and competence implies knowing the limits of your knowledge and skill. Here we have experts in conveying ideas and emotions—producers of television programs—violating proven rules of their profession because they confuse literacy with facility. These technologically astute people oversold technology to themselves. Their failure is more troublesome than a technically ignorant manager making technical decisions. At least in the latter case you have some hope that the technical water gets so deep so fast that the manager will call in help sooner rather than later. A little knowledge is a dangerous thing only when the possessors overestimate their intellectual prowess. Otherwise, it is a good thing when it allows you to carry on an intelligent conversation with people who can help you solve your problems.

3.5 WHERE DO I GO FROM HERE?

Right about now, you probably are saying, "Okay, so I need to know a lot of stuff about a lot of stuff. And I feel like I should have known it by now. But I do not. So where does that leave me?" Where most of your peers are, but in a better position than you were yesterday. You know that profits are what should get your attention, and you know that most people who work for you are looking for guidance, regardless of your management style. They will do your bidding if you stay true to yourself. (Those people who will not help you will have to be identified and unloaded, but more on that skill later.)

What you do not realize is that you already have most of the skills you need to be a good manager. They just have been overwhelmed by the myth machine of American business experts.

3.6 IMPROVING YOUR TECHNICAL LITERACY

So what is your exercise program to improve your technical literacy? First, honesty with your people and yourself, followed by a plan of action to fix the problem. To develop your action plan, write down your blind spots when it comes to key technologies in your company. For example, suppose you do not know how to run a lathe, but you have 16 of them working 20 hours a day. Honesty really is the best policy in this case because everyone knows you do not know your ear from a hole in the ground when it comes to these machines, so why not admit it out loud and get on with learning the language? By admitting the obvious, you replace fear and resentment with knowledge and the respect of those around you. There will always be someone who will gloat, but, by and large, you will be removing an issue standing between you and success.

You will find your people are eager to show you what they know; if they are not, find some who are. Got a union to work around? Now you are making excuses. The shop steward knows you are not training to take someone's job away. If you cannot get the shop steward to help you, your union problems are going to kill you anyway. One approach might be to ask a lead machinist to put together a simple program to bring you up to speed on simple, routine tasks, or you could take a correspondence course or two on machine tool design.

You say you do not have time for this? Have we got news for you. The practical, technical learning of management *never* ends. It is a part of your job. It is part of the job no one told you about. It is the part that, if you had a job description, would fall under the topic "Training." If you do not have time for learning more about your base technology, you do not have time for your job. So figure out what the lowest-value activity is that you do today, stop doing it, and replace it with real-value activities. What about going out to the shop or into the lab or perusing a book on applied technology instead of reading the *Wall Street Journal* or *Barron's*? As we will show you later, all you get from most business magazines are fashion ideas, not food for thought.

Chapter 4

Change Dynamics

Alexander the Great conquered much of the known western world in 10 years. Nearly one-quarter of the population of Europe died during the first Black Plague, which lasted four years. The great wealth machine of the Yankee seafaring towns lasted a generation, the Pony Express less than two years. World War II lasted about five years after the United States declared war on Japan. Each generation living during those times was convinced that the "rate of change" was the greatest the world had seen. Today's popular business press claims that today's business leader has to manage change like no other generation before. A quick review of U.S. business history—or any history, for that matter—shows that the presumption of "an era of change like no other" is a false but comforting myth. Believing that your generation alone is facing a unique challenge is a hearty bowl of comfort when you are on a diet of misfortune. It is easier to assign your failures to living during extraordinary times requiring extraordinary effort and luck, rather than to poor preparation, planning, and training.

Your time is no different from those that came before you, nor will it be much different for those that follow. The rate of change in human society has not varied much. What you use as your yardstick for change today differs from what you would use if you lived during the 12th century, but disruptions in the patterns of life (and, therefore, in business) come often and create opportunities essential to progress. Paradoxically, the great power of the human mind is that it incessantly seeks to find patterns in the chaos of life, many times finding patterns where none actually exist.

4.1 THE MYTH OF STABILITY AND THE REALITY OF CHANGE

The idea that life—and business—was somehow more stable in the past than it is today is a fallacy. Not only is it untrue, the fact is that the last 40 years of Western experience has been *more* stable than most previous eras. But stability is not the normal state for human endeavor. So if it is been relatively quiet for two generations, you can bet that all hell is about to break loose. Statisticians would say we are about ready to revert to the norm. We just are not sure exactly when. We are sure that when it happens everyone will claim we live in extraordinary times.

The good news is that history shows that people are very, very good at adapting to change. Not that everyone will come out on top, but over time humans always rediscover how to thrive in a naturally chaotic world. What is important to you is how to ensure that you are one of the survivors. Several things separate the survivors from the failures. Many people fret that luck plays an unfair role in success and failure. As one western-movie bad guy said (to his unfortunate victim before pulling the trigger), "Fairness ain't got nothin' to do with it." Luck by definition is available in equal proportions to all of us, but over time one thing stands out: Survivors capitalize on their luck by using their heads, just as those who survive battle are known for "keeping their heads." Sure, some of us are going to get through on the back of Lady Luck. Others, even though they have prepared themselves well, will be dealt a bad hand and get wasted. Those are the exceptions.

The problem is not that exceptions occur, but that some people believe that exceptions represent the norm. A few people around them get undeservedly lucky. Others seem to be fortunate to have "natural" abilities to solve problems. Convinced that these are the only two modes of success, those temporarily out of luck either wait for a hand to rescue them or delude themselves into thinking that their natural abilities will somehow, someday kick in. For example, how many managers pay lip service to planning when they really believe that luck overwhelms all planning? How many managers depend on their "well-developed" ability to judge character but are barely functional in that area? How many have said that they are good negotiators but cannot explain the basics? Many people delude themselves into thinking that any success they have is a result of a skill they do not have and their failures a result of forces outside their control. In reality, their successes and failures *are* outside their control because they have not tried to control them. They have not taken the time to think about how the business world is put together. They see patterns where none exist and miss patterns where they are unmistakable, and they get constant doses of rationalization from the business press selling myths as fact.

4.2 THE THIRD FUNDAMENTAL SKILL

It is no coincidence that we chose to follow the second fundamental, technical literacy, with the fundamental of change dynamics nor that we followed profit fixation with technical literacy. Not only do the six fundamentals complement each other to fashion a whole cloth of skills, there is some logic in the way they are rolled out for you. We started with profit fixation as the cardinal fundamental because the profit motive is the starting point for all business. We moved on to technical literacy because if you assume profit you must likewise assume an enabling technology that is the core economic value of any business, even if it is division of labor, the most rudimentary business technology.

Now comes the third fundamental, an appreciation of change dynamics. In Chapter 3, we introduced the idea of change by pointing out that what passes for high technology depends on your frame of reference. A frame of reference presupposes a system that is undergoing change. Therefore, if we claim technical literacy as a fundamental management skill, we must follow with understanding change, since the essential nature of both the business environment and the environment at large is something we call *change*.

If change is the natural state of being, and if the historical record shows that those who respond creatively to change wind up on top, then controlling the direction, pace, and impact of internal change must be a fundamental skill of general managers. But to turn the hackneyed saying "Change is the only thing that stays the same" into something more meaningful requires that you develop some sense about how all change occurs. In other words, you have to see a pattern within change. Once you see the common elements of change, you can do something purposeful when trying to get people to change in response to a changing environment.

4.3 THE THREE COMMON PHASES OF SUCCESSFUL CHANGE

You could fill a library with all the books that have been written about the change process. However, what you really need is a simple model of how people change so you can influence the direction and speed at which this naturally occurring phenomenon happens. In later chapters we will discuss how the change model is applied to business problems. For now, we need to understand change dynamics in general.

People who study human behavior have identified three distinct phases necessary to modify behavior successfully: unfreezing, changing, and refreezing. The first phase, unfreezing, is the most challenging one and demands most of your creativity as a leader. Unfreezing is the process of convincing those around you that staying with the status quo threatens their survival. The second phase, changing, is that amorphous state of change that always threatens to revert back to the status quo. It requires consistency, follow-through, and vigilance on your part. The last phase, refreezing, provides a neat reference end point. It is the marker that says the change is complete, that the sought-after modification has been completed. However, refreezing is an invisible point in time. Since the group has accepted the changed state as the status quo, it cannot identify the day the change was complete. Refreezing happens without most players feeling it. It is the absence of refreezing that is always felt. If the altered behavior is not accepted as the status quo, you wind up with a disjointed group, some wanting change, others sliding back to the old habits. Most failures to refreeze can be traced back to incomplete unfreezing.

Many of the business management fads built around controlling change emphasize technique over substance. They have elaborate management exercises, complex flow charts, and nifty jargon. In a later chapter we, too, will describe our favorite techniques and tools. However, developing the ability to recognize the common phases of change is essential if you are going to use those tools effectively. Since each situation you face will offer different opportunities for change and will present resistance in different forms, you need to recognize opportunity and resistance. About 80% of controlling change requires this facility. The remaining 20% comes from how well you wield the tactical tools.

4.4 SPECIALISTS USE SPECIAL LANGUAGE

Before we explore the elements of each of the three phases of change, we must clear up two things. First, we need to emphasize why a specialist language—*unfreezing, changing, refreezing,* and all the accompanying terms, such as *disconfirming information, change agent,* and *force fields*—is necessary. Second, we want to explore the first trap that you may fall into when you try out your new skills.

When you learn a new vocational skill, you expect to learn a new specialist language. Learning the lingo has several advantages. First, it makes it easier to communicate complex ideas to others who are also familiar with the terms. Second, it allows you to embody in a word or two previously unknown or unseen ideas. In other words, jargon gives form to the formless, potentially improving communication. Third, jargon identifies practitioners to each other.

There are many pitfalls to jargon, the most obvious is that many users never learn the deeper meaning of the concepts behind the words. Many of today's business fads—and change dynamics is surely one of them—are the consequence of superficial understanding of complex ideas. For instance, *change agent* is a term commonly used in change literature. Ostensibly, the agent is a person who guides, manages, and otherwise controls the change process. But if you do not go beyond the introduction and build on your elemental knowledge, you will always think that a change agent is a unique person occupying a special place in the group process. In fact, current business fad practitioners even offer courses that supposedly teach you to become a change agent, as if by assignment you, too, can be anointed. But as soon as you start using the term *change agent* as shorthand for this important protagonist, you shut yourself off from seeing a change agent as something other than an individual. A change agent is not necessarily embodied continuously in the same person, is very seldom appointed, and many times is hard to identify. However, since the word *agent* comes from our language of everyday use, the ordinary definition can overwhelm the special meaning.

For instance, in baseball one of the first things promising young batters learn is to distinguish different fastballs. The task is made easier because the only thing ballplayers see when first learning batting skills are fastballs. Young pitchers cannot throw much else with control, so they depend on "smoke" to get by batters. But as the batters mature, the word *fastball* takes on a more robust meaning. They no longer see those blurs they saw in their early years. Now, they see all the nuances: the different pitchers' preparation, delivery, ball movement, placement, and speed. The jargon becomes more meaningful, not less. Unfortunately, in the business world many managers get only a superficial treatment of many of their survival skills, do not practice those skills incessantly as sports professionals do, and just use them to impress themselves. When used often, meaningfully, and with care, jargon plays a critical role in the learning of any complex skill.

4.5 THE FALLACY OF MANAGING CHANGE

While it seems that the first trap to avoid when directing change is falling victim to the siren song of jargon, there is a more deadly one waiting for you. It is the idea that what you are doing is "controlling" change or "managing" change in others. It is easy to forget that you are the object of change, too. You easily can be seduced into assigning yourself the role of change agent. Many managers we have run into over the years will talk your ear off when describing how hard it is to change their people, but they fall silent when asked how they go about changing themselves.

So when you put some of the ideas presented here to work, do not forget to put yourself into the picture. Refrain from appointing yourself as the change agent. You might have influence but not because of any appointment. More important, if you do not participate in the change, those who work around you will see your efforts as nothing but managerial manipulation.

4.6 THE PARADOX OF FEAR AND SAFETY

All real change starts with overwhelming disconfirming information. People are not going to change the way they work, react, or think about a situation unless they are convinced that their current work, reaction, or thinking is hopelessly flawed, which is what disconfirming information does. Disconfirming information challenges your basic belief that things are okay. It is the information that says everything is *not* okay with your world.

For instance, why do you suppose advertising is so effective? Because it delivers the disconfirming message that your current behavior is somehow flawed. When you are being sold an inexpensive item, say a soft drink or a beer, building disconfirming information is not much of a task for the advertiser because the cure to your "poor" behavior is immediately available and relatively painless. You can buy a sports drink and shoot hoops like Michael Jordan or Shaquille O'Neal, or you can buy the beer and hang with the beautiful people. Advertisers seek to change your behavior, and they often use disconfirming information to build fear and thereby motivate change.

The problem with relying on fear to motivate becomes obvious when the risks are larger than buying a beverage. The larger the risk a person is asked to take, the more obvious and overwhelming the disconfirming information must be. But to endure such psychic pain, you must have a safety net, a handy cure. Suppose we tell you that your industry will be essentially wiped out in five years, that industry employment will fall nearly 40 percent during that time, and that you had better start looking for another line of work. (A real-world example of this scenario is the U.S. domestic oil industry between 1984 and 1989.) Not only would the evidence have to be overwhelming for you to believe it, you would have to convince yourself that the opportunities available in other industries are better than the perceived threat. In other words, for you to make such a large and risky change, you have to possess *psychological safety*. Otherwise, you will build all sorts of specious arguments against any evidence for change we can bring you.

In their small world, advertisers bundle disconfirming information and psychological safety together. They tell you that you are a nerd, but, hey, everyone else used to be just like you before they used the advertiser's product. They make it easy for you to identify yourself in the ad, and they make the solution seem effortless. This bundling of disconfirming information and safety shows up with a vengeance when sellers of fashionable business fads come to town.

This subtle paradox—that fear induced by disconfirming information can motivate only if it is accompanied by something that insulates the fearful—is why unfreezing is so difficult. The disconfirming evidence has to overwhelm because we usually are talking about significant change, but the vehicle for change has to be readily available to those who might feel threatened. This second element in the unfreezing phase of change, safety, is the one most managers cannot deliver unless they possess the six fundamental skills we introduced in the first chapter. Why? Because when facing a true, overwhelming, and undeniable need to change, people will look to their leadership for that safety. If the person in charge is confident and consistently delivers the goods, then those who look for that leadership will find the safety they need. We are not talking about leadership embolden by raw confidence. That is a given. We are talking about seasoning that raw confidence with a successful track record that comes only from possessing sustainable core management skills.

4.7 RECOGNIZING UNFREEZING

At this point, you know a few things you can do to direct change, whether or not your name appears in the "right" place on the organization chart. If your view of an impending threat is to be taken seriously, you have to provide credible, constant information about the situation as you see it. You have to offer (1) a working hypothesis for change, (2) shelter for those who might be threatened, and (3) a way for them to move ahead. There is a true story told of a small group of men isolated and trapped on the beaches of Normandy in 1944. This band of survivors had been separated from their assigned units. All their officers and NCOs had been killed or wounded. Organized leadership and communications had been obliterated. In the desperate huddle of men stalled on the beach, one rose to the occasion. Paraphrasing his words, he said that at the end of the day there were going to be only two types of people left on that beach—the ones who had already died and those who were going to—and he had no intention of joining them. He said he was going forward and anyone who wanted to live had no choice but to join him. While opportunities of such overwhelming fear, grave consequence, and obvious safety are unavailable to us in business, the lessons are available to us.

How can you tell if someone or some group is unfrozen? Simple—they begin to listen. You do not need special training to know whether someone is listening, but you do need to be in the same room, which has unfortunate implications for those who chant the "Technology-will-replace-human-contact" mantra. This leads us to the next reason business change initiatives fail: The person who has the authority to commit funds relies on reports and second-person accounts, never seeking out those who are the objects of the change. Without good, firsthand information about how the change is progressing, the responsible manager distributes rewards and punishment haphazardly, dooming the project.

One of the truisms about engineering is that not everything you need to know to manufacture an item can be put on a print. The same can be said of this model of change. You have been introduced to the three phases of change and have been given some details about the first phase, unfreezing. But do not fall for the idea that this is a linear activity, even though the model implies that it is. People do not reach an "unfrozen" state and stay there waiting for someone to come along and manipulate them into the second change phase. You have to continually paint and repaint your landscapes entitled "The Failure of the Old View" and "The Triumph of the New View." Otherwise, people will naturally slip back into the old state.

4.8 THE FALLACY OF EMPOWERMENT

Later we will look at different management fads in detail, but this is a good place to introduce a criticism about one in particular: empowerment. Many people believe that empowered subordinates can find their own self-defined problems and generate their own solutions. The assumption is that people, given the green light by management and the right tools, are the best device to initiate, implement, and resolve change. Management's role is to get out of the way of this natural creative energy. By this time, you must be convinced that while that premise sounds nice, change does not happen this way.

Change is something that happens to people, not to inanimate objects like "the company" or "the department." Changing people's ideas, focus, and energy is a leadership responsibility. Lacking such leadership, people targeted for change will not be able to understand, relate to, nor care very much about the new attitudes, assumptions, or work necessary for change. Not that they actively resist changing, but without a unified vision they cannot sustain change. No one can deny that the opportunity for change, the energy required to move forward, and the identification of alternatives can come from individual group members. But it is beyond comprehension that a business fad such as empowerment can assume that useful, profitable change will happen without intervention, without a plan, and without recognition of how people and their social groups actually change.

4.9 REFREEZING

While unfreezing is still fresh on your mind, let's discuss its converse, refreezing, which is not the next phase but the third and last phase of change. The second phase, the state of change itself, is actually the shortest and easiest to work, given that the targeted people are unfrozen and ready for change. Because the second phase is the most tangible of the three, and because it is the one that most people identify as the sole artifact of change, the literature is full of advice on how to "manage" change. We, too, will contribute a few of our favorite change tools, such as free-form organization charts and force-field analysis, for you to use while your change project is underway. But for now, detailing those neat accessories would be fashionably distracting rather than informative.

The most difficult concept to accept about change is that refreezing is not something you do. We can identify activities that help to unfreeze a work group: collecting and sharing disconfirming information; identifying passive resisters; looking for opportunities to suggest safe passages to the future state. We can also identify activities that are helpful during the change process: revisiting the reason for the change; deciding how progress will be measured; drawing force fields to illustrate the forces working for and against the change; identifying players who are so disruptive that they have to be removed. But refreezing is not an activity-based change phase. It exists in the collective minds of the players. For example, take the typical experience of smokers who successfully stop smoking. While they may be able to tell you the date they had their last cigarette, they will not be able to tell you the date they changed from being an ex-smoker to a being a nonsmoker. The day they stopped smoking was the start, not the end, of their effort to change. The end of the process is less recognizable. Nothing changed on that date, except everything.

4.10 APPRECIATING THE PROBLEM

A good way to appreciate how hard it is to change and what the different phases of change look and feel like is to try a controlled change experiment on yourself. First, pick some innocuous habit you want to change. Avoid the typical hard ones like losing weight, cutting back on alcohol, learning a foreign language, or appreciating your in-laws. Those change initiatives require so much disconfirming data and huge amounts of psychological safety that they rarely are done without help from someone else or completed in a short period of time, nor is the refreezing phase very distinct. Instead choose something you would like to do differently but will not take much more effort than motivating yourself—or so you think. Some typical, low-risk change initiatives follow:

- Putting everything in the family room away before you go to bed;
- Closing cabinet drawers and doors when you are finished with them;
- Getting up when the alarm goes off—consistently;
- Getting regular, low-stress exercise, like walking every day after work;
- Asking for the manager each time you get poor service;
- Taking time to read each day;
- Generating a daily to-do list before you start your work for the day;
- Starting every meeting with the word *profit;*
- Taking notes during every business meeting, no matter how short;
- Ensuring that you talk to each one of your direct reports each day.

Once you have chosen the change initiative you will use to demonstrate the change process, keep a simple, single-line, one-page diary about your attempts to change your behavior. It will be easier if you pick the same time of day to update your diary—before you leave work for home, before you go to bed—so you do not forget. Note that keeping a diary is a change in itself. Keep these notes for at least a month.

What you will find will be fairly typical. First, when you feel like backsliding, you will question whether the change you want is so important anyway. On certain days, you will rationalize why you just could not work on your project. Many people will give up after the first week of this challenge without giving up, which is okay. You need to know how hard it is to change behavior, even when the risks are minimal.

But if you are successful in changing some part of your life, even if it was not the change target you set for yourself at the beginning, you will notice that a vital awareness developed during the change project. At some point you were able to sustain the change—refreeze—because the value of the change became so apparent and overwhelming that you could not imagine going back to your old behavior. If during your forced march you fail to reach that state of awareness, you most likely will be unable to make the change permanent. On the other hand, if you changed a bad habit common to other folks, too, you will find that you will become more enthusiastic about the possibilities of change and will want others to experience the change themselves. This pseudo-religious conversion experienced during your personal change project has implications for changes in your company. Someone has to get religion in order to generate enough energy to sustain the change initiative to the refreezing phase. If you were lucky enough to see that happen in the context of your personal change, you will find it easier to recognize the conversion when you are working on group changes.

In the end, this little exercise is useful regardless how it turns out. If you fail, you will understand how this change model can identify the problems of directing change. If you succeed, you will appreciate the intense satisfaction you can get from working through a seemingly simple but ultimately complex change.

4.11 THE MYTH OF MODELS

We mentioned that the change phases we presented are a model of change. Models are used to explain things that are difficult to visualize. A word of caution, however, about business models: They are not reality, nor is their predictive output *in itself* very valuable. This significant distinction will keep you out of untold mischief and is one we will come back to more than once. We use models all the time in business. Your financial books, product design procedures, and sampling inspection programs all depend on models that simplify reality. The change model presented in this chapter is obviously a model of individual and group behavior.

One of the myths of models is that their predictive power is their predominate value, when in fact their real value is diagnosis. Take engineering and science, disciplines that have elevated models to near godlike status because of their ability to predict behavior of physical systems. That status results from pedantic methods of instruction that destroy understanding in favor of facility. Being able to calculate a differential is rewarded more than being able to recite its useful application. For example, most engineers claim that, given appropriate tools and background, their skills will allow them to predict product performance and reliability, which is undoubtedly true. They send reports to management showcasing their sophisticated models and predicting a certain capability and capacity for a proposed product. What they leave out of their neat recollection of their design is how they actually used the model. While they used it as a starting place for the design, they also used it to diagnose significant problems that always emerge during product testing. Engineers are reluctant to admit that their models are mere recollections of test data and field performance gathered over long periods of time. They would rather believe that models are a testament to engineering finesse and creative insight. Not that engineers are dishonest; it is just that their dominate paradigm classifies output as a model's predominate value.

The same should be said of this model of change dynamics. Like all models, it should be classified as a diagnostic tool rather than a predictive one, even though it does both. While it might help predict what the next state of change will be, it is best used to figure out what the problem with your change initiative might be rather than forecasting where problems might occur in the future. When a product unexpectedly fails during testing, engineers go back to their models to identify heretofore unrealized but now identifiable forces. When a change initiative gets into trouble, you need to go back to your model of change and see where the problem might be. Prediction is fun and must to be done to set up the initial design or experiment. It is a mistake, however, to miss the diagnostic value of models in preference to their predictive value. As you will see in later chapters, precise predictions seldom come true, no matter what their source or apparent sophistication. When they do pan out, the predominate factor is luck, although most owners of successful predictions will point to their superior but always proprietary model. Using models for predictive purposes is a parlor game compared to using them for diagnosis. Diagnosis is much harder work, more emotionally intense, and abundantly rewarding because only diagnosis can be used to improve mediocre or failed performance.

4.12 GOALS AND OBJECTIVES: WHAT IS THE BIG DEAL?

While having a model that explains change is useful for understanding the dynamics of change in general, what practicing managers really need is to translate this generalized model into practical ones that give them a toehold on the mountain of problems they face in the real business world.

Even before embarking on a change initiative, a manager needs to answer two elemental questions: "Why change?" and "What are we going to change?" Finding satisfying answers to those questions is not as easy as it seems. Counterintuitively, you should answer "What?" before you answer "Why?" The logic to this approach is inescapable. You will not know if a certain change should take place if you have not adequately explained what you are going to change. Much like scientists, who first must form a hypothesis before they can figure out what kind of experiment needs to be built to test it, you have to start with an idea of what you think needs to be changed before building your "experiment" to test whether the change is necessary. As it turns out, explaining what you think needs changing is actually easier than determining whether or not to change.

A common mistake is to answer what should changed by stating goals rather than objectives. We regularly encounter managers who claim that an upcoming project will "break down barriers," "improve communications," "strengthen morale," "improve quality," "change the way we think," "help us become more entrepreneurial," or "empower people." While all these change objectives regularly appear in the general business press, they are goals, not objectives. What is the difference between the two? None vernacularly, but a great deal in the jargon of business. A goal is the end toward which one moves, while objectives are those things, usually physical, that one must complete on the way to reaching a goal. In other words, to reach a goal, we must achieve a series of objectives. The difference between the two intentions can be remembered from their etymology. *Goal* comes from the Middle English noun *gol*, a boundary or limit; *objective* from the Latin verb *obicere*, meaning to throw in the way or hinder.

By now, it should be clear that you cannot design a change project to achieve a goal. You have to translate your goal into more concrete, measurable objectives. The difference is obvious when you try to define a project to the people who will be involved. For example, take the goal of improving communications. That is one you hear often after a train jumps its track. "We've got to improve communications so screw-ups like this do not happen," says the boss. The postmortem meeting usually breaks up with renewed commitments to keep each other "more informed." A week later, nothing has changed. When we say nothing has changed, we mean work habits remain unchanged. You still hold the same meetings in the same way; you still issue and receive the same reports; you still report to the same person and the same people report to you; your daily routine continues unaffected. If you had stopped to say, "Keeping each other more informed is a good goal, but how are we going to achieve that?" you would have framed the right question. Your change initiative can now be explained in concrete terms. That is the only way you can develop alternatives for action.

Incidentally, whether you intentionally do so or not, those alternatives will always be directed at changing the way work is done. Real change inevitably involves the *system*. In the case of trying to "improve communications," maybe you will design more structured staff meetings; overhaul your nonconformance reporting system; change work assignments; redesign reports; reassign people; institute more results-oriented training; hire someone; or fire someone. In any case, your change objectives result in tangible changes to work routines.

4.13 CHANGING TO IMPROVE PROFITS VERSUS CHANGING TO IMPROVE LIFE

So it comes down to this: The objective of any change initiative in business is the method of work. In the trenches, change is not about improving communications, although better communications may result. It is not about improving quality of work life, although people may enjoy their work more. It is not about strengthening morale, although people may show more esprit de corps. Changing a business means changing how people do their daily work, which, in turn, may affect the way they look at their work and the people around them. However, that is a consequence of pursuing a change objective; it cannot be the objective itself. Changing work is how profits are increased and expenses reduced. If you have a change project with an objective to change anything other than work patterns or output, you are wasting money.

That is not to say that scheduled sales-indoctrination meetings, complete with pep talks from the vice president of sales and maybe an outside motivational speaker, is not a good change tool. However, such change activities have to be seen for what they are: training. What we have to remember but often forget is that such meetings must have an outward, measurable result, a change in the way work is actually performed, which results in increased profitability. If the activities do not yield measurable results, then the change initiative is an illusion in which activity displaces achievement.

In summary, the objective of any change project in your firm has to be a change in the way people in your organization do their work. All other outputs are irrelevant.

4.14 WHO SAYS WE HAVE TO CHANGE ANYWAY?

The first change phase, unfreezing, in practice overlies the second phase, the state of change itself, because we find ourselves asking the first-phase question "Why change?" throughout the change process. Forcing the question ensures that we remain thawed out. Therefore, we have to ask it often and answer it vigorously. In the beginning, it is fairly easy to answer with a list of compeling reasons. As more information about the actual change activities becomes available, the risks and rewards of the potential change are better defined. As you pursue your objective, you reach a point of maximum doubt whether the project is viable. One of the most debilitating management tendencies is to invest time, effort, and money in change initiatives that are premature or simply not appropriate for the company. They get caught in this trap because they do not ask the "Why change?" question enough or answer it disingenuously. The result is a never-ending parade of new change initiatives, started with fanfare and deep commitment only to die a slow, unnoticed death.

In the beginning of an anticipated change initiative, the reason for changing work habits may be easy to state, and the answer to "Why change?" is treated as obvious. But only after you generate alternative action plans can you appreciate the number of people involved, what ingrained habits have to be changed, who and how many people are aligned against the change, and how much safety can be constructed for those who will be affected. Only after you have answered the question "What are we going to change?" are you in any shape to answer the question "Why change?" in any cogent way.

Because the order of these two start-up questions actually occurs out of "natural" order, many general managers get caught up in change initiatives that should have been killed. It is difficult to stop a process with a question you already have answered or one that was treated as obvious. It is not a question you would generally ask unless you appreciate that unfreezing is a management activity that has to be continuously applied throughout the change initiative.

4.15 WHAT DOES "YES" SOUND LIKE?

As with most real life questions, your answer to, "Is this change necessary?" becomes more equivocal, tentative, and conditional the more you ponder it. How do you know whether you should proceed with the change project, knowing that the answer to the most critical question will be, for the most part, wishy-washy? There are several obvious signs. First, while your answer always becomes more tentative and conditional over time, the conditionality should become more concrete and manageable, and your anxiety about the project's righteousness should decrease. If you do not see that trend, you probably should abandon the change project. Second, even if you classify the change as mission critical, you may find that the action items are hopelessly incompatible with the assets you have. In that case, you have to abandon the project.

Canceling a project is as important an activity as proclaiming a new one. It demands that the company leadership be honest about what is possible and what is not. In some cases, the desirable change is beyond the capability and capacity of the firm, in which case the original change project may have to be abandoned in favor of a more fundamental change.

For example, suppose you want to design and install a documented standards-based management system in your company. You decide that if you do not your overhead and rework costs will continue to grow. Your survival in an increasingly price-competitive market is threatened unless you get a handle on those costs. This is a mission-critical project. As time goes on, you realize that you have had little success convincing your staff that this project holds the salvation of the company. Worse, you realize that you yourself are having trouble adjusting to emerging price competition in a market that has been, until now, more focused on product features than price. The longer you attempt to make a change that neither you nor your staff feels compeled to make, the less time you have for developing alternatives that may work better for both of you. There are other genuine, understandable, and supportable reasons for canceling a change initiative. What was first seen as critical becomes less so when more information becomes available. Priorities change. Other opportunities present themselves, and, given limited budgets and personnel, you may find it necessary to shift to other work.

4.16 KILL IT BEFORE IT DIES OF NEGLECT

Companies abandon mission-critical projects all the time. It is just that management usually avoids public burials because they are more concerned with saving face than saving money. Instead of presiding over a very public burial, most hope the failed projects will just fade away. After a while, those managers ensure a self-fulfilling prophesy. The disregarded programs do seem to fade away. But what those managers do not realize is that the programs have instead transmuted into profit-eating ghosts roaming the halls. Over time, the office and plant will be full of the walking dead. Management pretends not to see the zombies, or, worse, they see them but hope no one else does.

Why do such projects turn into ghosts rather than just die of neglect? First, if you do not give an abandoned project a public burial, people will generate their own epitaphs. For example, some will say management is so unsure of themselves they cannot admit to a mistake; others will be more charitable and say that the project was a mistake. Of course, attempting a change is never a mistake; not giving it a necessary burial is. Since you did not take the time to explain to those who worked on it why the project had to be abandoned, your neglect turns what was a good fight into a feeble fuss.

The second danger in avoiding public burials is that people will begin to disregard and even avoid all new projects. They begin to see new projects as a manifestation of management ego rather than real attempts to solve real problems. The ghosts of abandoned projects wander the halls, adding more fuel to the fire of office rumors and innuendo.

Last, without a clear signal that a cherished project has been canceled, some people will continue to work on it way beyond the time they should. Instead, they proudly generate valueless work. Just because you assign new work or tacitly encourage them to return to their old work without clearly rescinding the previous objectives, some people will not see that as a clear signal. In fact, in any group of people attempting a change, there will be those who, for various reasons, will want to hang onto the project until the bitter end. It is the first time in a long time they think they have had something worth contributing. But the longer you put off the inevitable, the worse they are going to feel about the cancelation of their favorite work.

4.17 SOLVING THE PROBLEM

You undoubtedly have several ongoing change initiatives at your work. Whether or not you see yourself as responsible for them, write down what you believe their work objectives are. Ask the question, "Are these changes necessary?" Write down the various reasons you believe the projects should continue. Then put the paper away for six weeks.

In six weeks, look at what you wrote. Compare your thoughts six weeks ago with what you believe today. Have the projects really moved forward? Are they more pressing than they were before, or less? Have new ones replaced old ones? Have any been completed? Have any been put out of their misery? Look at the entire list of change initiatives within your company, not just your pet project. Do they really result in change, or are they just change for the sake of change?

Another benefit from this little test is realizing the power of paper and pencil (or the computer equivalent). We are not talking about generating memos (or e-mail) on a whim. We are talking about the craft of reducing ideas to paper. Too many companies either do not write enough down or write too much of the wrong thing. The most important management information you can write is an explanation of your decisions. This historical information is critical to good management because it allows you to revisit decisions without depending on your memory. While the human mind is without equal, the astonishing thing about it is how unreliable one's memory is. If you are paid for the output of your gray matter, the most important thing to remember is that you do not remember.

For example, have you ever gotten in an argument with your significant other in which both of you were adamant that your recollection of a shared event was the right one but the two versions were 180 degrees opposite. Not only are the recollections different, but both of you earnestly believe the other is totally and unmistakably wrong. The fact is both of you are telling the truth. That is why eyewitnesses to a crime are notorious for giving conflicting accounts of what they saw even when they were standing next to each other. The power of the mind to synthesize coherent pictures of life from small bits of recalled information is what gives the mind its power, but it is also the source of self-delusion in business. The bottom line, on the other hand, rarely forgets.

What you need to do is prove to yourself that not only do projects become the walking dead but that your recollection of the facts and circumstances surrounding their past is unreliable. With that proof behind you, you will appreciate the need to use a planned and documented record to keep track of every change project. That record will give you the undeniable information you need to cancel some programs and encourage others. Without the facts, you live a fantasy—and while fantasies may be fun, they are seldom profitable.

Chapter 5

Useful Change Tools

After you get a handle on the three change phases and try a simple personal change on yourself, you begin to understand why the failure rate for corporate change initiatives approaches 70%. But appreciating the reasons for the high failure rate of those programs is not the same as finding a remedy. Improving the success rate means we have to change the way we envision, devise, and implement planned change. That means we have to do something radically different during the second phase of change, the implementation or transition phase. So far, we have discussed the things you need to do in the unfreezing phase of a company project. Now, we look at the things you need to do to improve the chances that your process for change stays on track during the long, hot summer of change, when new ideas seem worn and everyone gets a little drowsy.

But before we get to the nuts and bolts of managing the transition phase, we need to emphasize that the change skills outlined here are good for all kinds of change projects, not just the latest business fad. Not only do "new" companywide change initiatives (like TQM in the 1990s) fail at unacceptable rates, the failure rates of all "special" projects are historically and inherently high. Recently, we attended an industry forum of software engineers concerned with out-of-control software development budgets and resulting mediocre product performance. Different speakers repeatedly quoted widely accepted failure rates of 70% in their industry, a failure rate strikingly familiar to professionals involved in high-profile corporate quality initiatives. The audience listed familiar reasons for failure: vague and constantly changing objectives, underestimating the resources and time required, inadequately trained personnel, and the ever common "poor communications." This last cause triggered someone to mention the problem of technically challenged management. Snickers filled the room, then embarrassment. All those reasons were generated within three minutes and, in some form or another and to some greater or lesser degree, are common to the failure of every botched change initiative. This commonality across industries and professions strongly suggests that responsible management lacks a basic understanding of how to keep projects on track and how to help people remain focused while implementing change, regardless of the technical disciplines involved.

5.1 THE USUAL TOOLS

People with experience running large or complex projects are familiar with Gantt charts and similar project management tools. While those tools are widely taught and in many cases invaluable when managing some projects, we are more interested in what they *cannot* do rather than what they can do. Like so many management tools, knowing how a tool like a Gantt chart can be misused is as important as knowing how it is used.

For those who need a reminder, Gantt charts (sometimes called time lines), are named for the man who popularized them. A Gantt chart, like the one in Figure 5.1, visually represents a project's tasks, the tasks' relationships with each other, and the projected and actual start and completion dates. The sophistication and versatility of these charts make them indispensable for managing complex, detail-driven projects, such as commercial building construction and civil improvements. With the advent of PC-based project management software, Gantt charts also can be used to model and track smaller projects, such as new product development or plant extensions, cost effectively. While project management tools usually are associated with activities that have physical outputs, like a building or a product, they frequently are used to control companywide change initiatives, such as reengineering and TQM as well.

The difficulty in using a time line to represent project progress is that *process* can be mistaken for *progress*. For instance, say you decide to install a documented standards-based management system like ISO 9000 or your industry's *good manufacturing practices* (GMP). Your project goal seems easy to state: "To obtain second- or third-party certification of our documented system." Early in your planning cycle, you reduce all the necessary action items to various tasks, assign responsibility, estimate required time, and use a canned computer program to determine the critical path, relationships between the various tasks, and the estimated completion date. You use the resulting Gantt chart as an early warning tool, allowing you to determine where you are diverging from your initial plan or, conversely, how wrong your initial plan was.

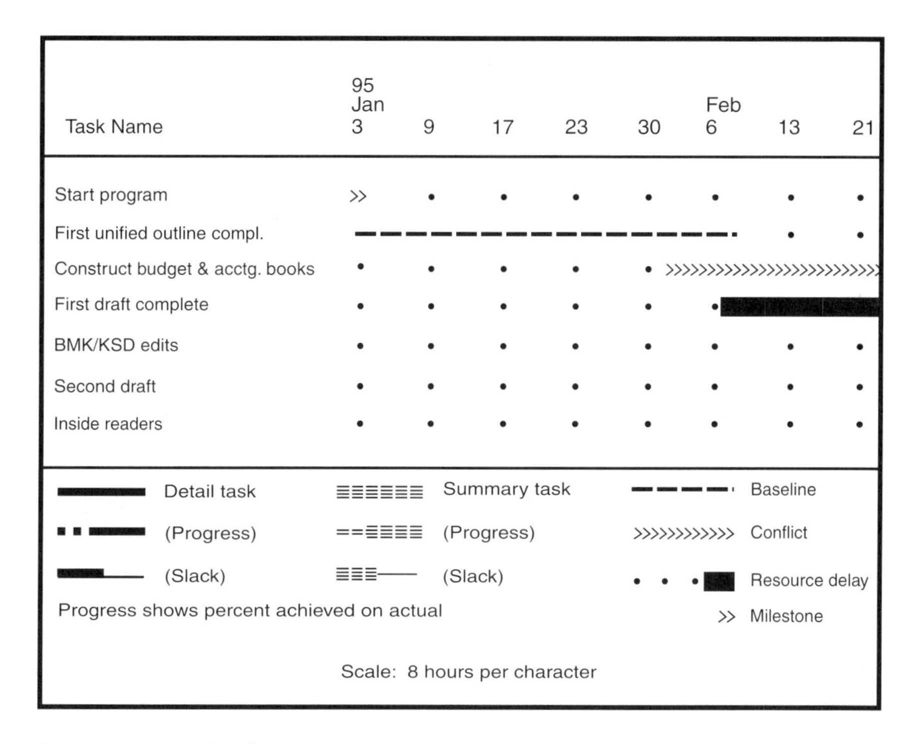

Figure 5.1 Example of a Gantt chart.

In Chapter 4, we cautioned against using models solely for their predictive power and neglecting their diagnostic power. The same must be said of Gantt charts. It is easy to appreciate their predictive power. After all, if a certain task takes longer than the time allotted, most PC-driven Gantt charts will automatically predict how such an early-stage problem will affect subsequent tasks and how far the finish date will move. Commonly overlooked is what those cascading adjustments imply about project health. Unfortunately, Gantt charts are not issued with a warning that states, "The root cause of changes needs to be investigated." Instead, the ease at which you can make changes to these charts insinuates that scheduling problems are routine and schedules flexible. While any particular scheduling problem might be routine, repetitive rescheduling indicates that the original assumptions concerning allocated resources and costs are suspect, or that the project is headed in the wrong direction, or that some other of the many common and systemic problems associated with planned change projects may exist.

This is another instance in which a model intended to illustrate deviance from original intentions becomes a pseudo-predictive model. What is forgotten in the thrill of the newly accessible technology is that the original schedule was a best guess; therefore, what later passes for prediction remains a guess. That type of misuse of models can be avoided if you train yourself to be less enthralled with their predictive powers and instead more appreciative of their diagnostic value.

Before we leave Gantt charts, we should emphasize that they, like all pseudo-predictive models, are good tools and do indeed have their place. They will unfailingly tell you the impact of lagging tasks on the whole program, assuming that the rest of the program follows your original best guess. In recurring programs, like building a highway or dam, institutional memory is a great asset. In those circumstances, Gantt charts are good tools to ensure that tomorrow's project leaders can capitalize on today's lessons. However, while their predictive power is central and obvious and improves with repeated use when applied on the same repetitive project, their diagnostic power, unfamiliar to many, can be applied to nonrepetitive projects. Therefore, diagnosis is the predominate value of any model when you are facing systemic problems like the overwhelming failure rate of corporate change initiatives. While you might find that a Gantt model is necessary to track all the interrelated tasks of a complex change initiative, that is seldom sufficient to ensure that projects stay on course. Even though they show progress against plan and imply where root causes might be hiding, they cannot illustrate the state of more fundamental problems like resistance, relationships, and momentum. You need a different set of project tools to illustrate the current state of change and to reinforce critical change behavior.

5.2 THE UNCOMMON BUT MORE USEFUL TOOLS

If Gantt charts are nice but overrated project management tools for the kinds of projects in which we are interested, what is left to help us manage change? Up to now, we have emphasized the need to define your change objectives in terms of work output, the importance of regularly revisiting the necessity for change, and the value of recognizing the three distinct phases of change. We have explored the first phase, unfreezing, at length because it is so essential to successful change. We mentioned that the third phase, refreezing, is merely an end point that we usually note in passing. The second phase of change, the transition itself, remains to be explored. While many change projects fail because of insufficient and incomplete unfreezing, what is left fails because we cannot sustain the energy level and awareness required to see the project through to the end. To that end, you need tools that will help you overcome what we know are the typical reasons change initiatives fail during the transition phase: loss of focus and loss of energy. First, we will sketch each tool, then go back and flesh each one out.

5.2.1 Understanding the Gestalt Cycle of Experience

Loss of focus is loss of awareness. If we are going to work on awareness, we (naturally) need to develop a model that explains "awareness." The model should explain what we mean when we say we are aware of something, how that awareness comes about, and, most important, how a group goes from the intensity of collective awareness during unfreezing to the feebleness of neglect when the project grinds to a halt. The Gestalt Cycle of Experience is an excellent model for that process as well as a surprisingly practical way of understanding the role awareness plays in change. By appreciating this seemingly simple behavioral model, you will understand how awareness and focus might be influenced, how to keep on track, and, more important, how to help other people avoid the inevitable dissipation of attention.

5.2.2 Running Good Meetings

While it is trendy to avoid meetings because they have become superfluous, significant change projects cannot sustain themselves without meetings. Meetings *are* superfluous time-wasters because managers do not know the basic elements of good meeting stewardship. You need a few unerring rules to get more value from a meeting, whether it be a chance encounter in the hall or more formal command performances.

5.2.3 Using Free-Form Organization Charts

Another hated, disused, misused tool is the organization chart. We need to understand how to use conventional charts and what their weaknesses are. From that basis, we introduce something we call *free-form* organization charts, which illustrate how work actually gets done. This new tool exposes the nature of the organization behind the organization, an understanding essential for removing or reducing barriers to change.

5.2.4 Applying Force-Field Analysis

All change is supported by forces encouraging the change and opposed by forces hostile to it. Force-field analysis is a handy way of illustrating which forces help and which hinder change. Such analysis provides a historical benchmark that is particularly helpful when you become frustrated at the slow pace of change. It also visually reinforces a truism about change, that most project breakthroughs occur when forces opposing change are removed rather than when supporting forces are invigorated.

5.2.5 Measuring Change

In business, if you cannot measure it, it does not exist. Unlike measuring changes in the physical world, measuring change in the business world requires creativity, not analytical prowess. To give you a sense of what we mean, we first will give an example of this kind of ingenuity in action. Then, we will provide a checklist to help you ensure that you concentrate on the more important measurement attributes when assessing the cost/benefit of a change.

5.3 FOREGROUND, BACKGROUND, AND GESTALT

According to our dictionary, *awareness* is "having or showing realization, perception, or knowledge" and "implies vigilance in observing or alertness in drawing inferences from what one experiences." All of us are aware that we are aware. We also are aware that another person's awareness differs from ours. Even so, we know that we can affect the awareness of others because we do it every day.

Suppose you get on a subway, and the person sitting next to you lights up a cigarette, contrary to the "No Smoking" signs. You mention to her that smoking is not allowed on the train. She promptly has a snit and calls you various uncharitable names. The next thing you know, everyone is razzing your neighbor about her cigarette. She puts it out and mumbles a few choice good-byes to you on her way out of the car.

Let us analyze what happened. It is likely that if you had not said anything about her behavior, no one else in the car would have spoken up, either because they were not bothered or they did not want to bother. By speaking up, you brought their awareness of a particular issue—rudeness—into their mental foregrounds. Apparently, many in the car, once you brought the issue up, realized that they really did not care for the smoker's rude behavior either. At that point, it is easy to believe that your fellow riders were waiting for someone else to say something. While some may have been, most were not. In a car of 30 or 40 people, the chances that perfect strangers are reacting to their mundane environment exactly the way you are is very, very slim. People get lost in their own personal mental foregrounds and backgrounds. Some are tired and content to daydream. They do not want to interact at all. Some are looking forward to seeing their family or friends. Some are still struggling with a problem back at work. Some are not feeling well and are concentrating on their discomfort. Once you brought up your issue, their mental foreground was disrupted, but each one still had a personal and specific reaction to the situation. That the crowd's intensity was equal to yours once its collective awareness caught up to yours was just coincidence. In another time and another place, they would have left you and your new friend to settle your differences.

This feature of our self-referential world—that our perceived world (and therefore the business world at large) exists only in the mental foregrounds and backgrounds of individual people—is a powerful tool for understanding how groups lose focus on change objectives and how you can get them back on track.

The preceding example is a layperson's application of the Gestalt Cycle of Experience to group dynamics. You probably use fragments of this model without realizing its origins. Have you ever said or heard it said that you have gotten (or reached) "closure," meaning that you have reached a consensus among different parties about a common issue and that each person was preparing to move on? *Closure* is part of the jargon of Gestalt psychologists, but others who use the term are unaware that they are borrowing from that field of study. Furthermore, Gestaltists say that closure also represents learning, meaning that the actors have extracted the meaning from the common experience.

Figure 5.2 is a simple illustration of the cycle for a single person, while Figure 5.3 represents the cycle for two people engaged in a common activity. These representations are far from being technically rigorous. We have left out many details, simplified definitions, and mixed various levels so we can extract the central value of the model for you: Successful communication of, and action toward, any goal requires that all parties reach peak action states at the same time. It helps when the intensity of action is equal among the parties, but a lack of intensity is not as counterproductive as people reaching the action plateau at different times, that is to say, be "off-cycle", as it is in Figure 5.3. The two people in that figure will have a hard time moving the project ahead because they are seldom on the same wavelength. When one is ready for action, the other is not. That inevitably leads to each believing the other has no interest in the project, when in fact they both do, but at different times and at different intensities.

The speed at which a project can be completed is related to how long the action phase can be sustained or how often the cycle repeats. If the two people in Figure 5.3 remain off-cycle, they may still get something done together if the cycle repeats itself often enough.

The action phase cannot be sustained indefinitely; closure has to occur in order to build understanding necessary to start the next action phase. If people do not get closure often enough, they get tense and dissatisfied with their progress. Everyone has to take time to look back at their experience before moving forward. If they are forced to move forward before they get closure, all the open issues crowd their foreground.

Change initiatives that become dysfunctional during the transition phase almost always do so because the responsible leader lacks an apprecia-

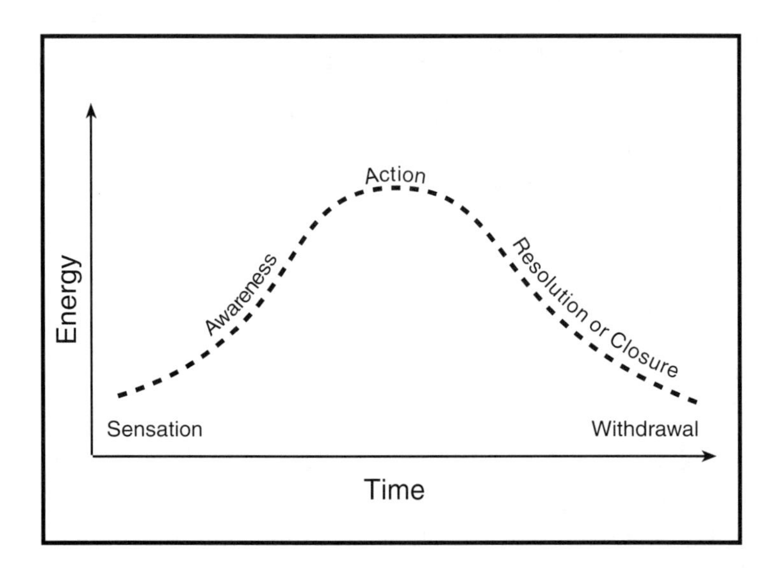

Figure 5.2 A simplified Gestalt Cycle of Experience.

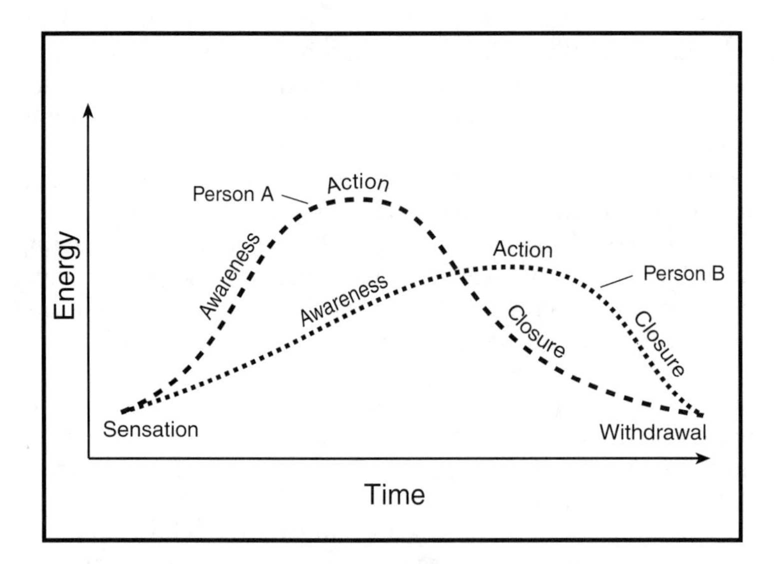

Figure 5.3 A simplified Gestalt Cycle for two people.

tion of his or her role as the instigator and supporter of the phases preceding and following action, that is, awareness and closure. While all the phases of the Gestalt Cycle are personal and internal activities, awareness and closure are the two where an adroit leader can be most affective. Some leaders have learned one-half of the lesson and are known for "keeping the pressure on." Their success with change is higher than most, but they would be more successful—and less disruptive—if they spent an equal amount of time on the closure needs of their people.

This simple model of interaction, although a complete and indivisible cycle, is centered around awareness. The issue of awareness is something we encounter every day yet hardly reflect on. Not only is the Gestalt Cycle a helpful tool for diagnosing project implementation problems, it is also extremely valuable for organizing your own thoughts about your personal problems with self-motivation. Awareness is the essence of the transition state. Only if you as the leader can see how you can affect individual awareness during the transition state can you bring energy to bear on the obstacles to change.

Let us look at a small change project. Say you want to teach your dog to retrieve a Frisbee. First, you have to build the dog's interest. Some dogs are not going to have a problem with that. Anything you are interested in, they are interested in. But suppose your pup is more discriminating. She enjoys playing with your old sock but shows no interest in the Frisbee. She just sits, looking at you play with that plastic disk and thinks that (like so many other things you do) you have lost your mind. Then one day she starts showing interest. The next day, she backslides. Nothing short of sticking the Frisbee in her face makes her the least bit intrigued. But she always has that old sock. At this point, you have to decide if you would be happy playing with her favorite sock or if you want to continue building interest in the Frisbee. You know that given enough time and patience, she will learn that the Frisbee is your preferred toy.

The point of this example is not that people are like dogs, but that if you want to reach your objective, you need to constantly monitor and reinforce awareness levels during change initiatives, both your awareness as well as others'. A particular problem with many managers entering the transition state is that they fail to realize that they alone are responsible for monitoring awareness. Too often, people know that they can passively resist a change just by ignoring it. They know that the boss historically shows only episodic interest in these types of projects. With brief, sharp awareness or action peaks and long lulls between cycles, they know that they can doom the change project. The critical factor for project leaders is that their own awareness has to precede awareness in others.

In the dog-training example, changing over to playing with a sock is no big deal if your objective is to enjoy your pooch. However, if your objective is to train a champion free-style Frisbee dog, you might retire your current dog to the country and get yourself another dog. Changing your goal to sock playing in the middle of the project is okay, as long as you do that consciously. You have only replaced one project with another in light of the resources and energy available. But if you change it because you let your foreground picture fade due to neglect, you have not a project but a problem. Similarly in business, if you start program after program as a reaction to letting what is foreground and background change without thought, then your company will be dysfunctional.

Now that you are aware of awareness, how do you ensure that awareness remains high and projects move forward? The core implementation activity that boosts awareness is well-run meetings.

5.4 ALL YOU REALLY NEED TO KNOW ABOUT MEETINGS

Our dictionary defines *to meet* both as "to encounter" and as "to come together with, especially at a particular time and place." Similarly, we are interested in both the short, two-person chance encounter in the hall as well as the typical conference-style meeting in which people gather to discuss predetermined issues. In both circumstances, at least one participant is attempting to influence the awareness of the others. In the realm of change initiatives, these meetings focus people's attention on the ongoing transition phase.

Meetings are useless if they fail to bring the group to common levels of awareness or if they fail to reinforce closure for all participants. Even though meetings seem to be universally disliked, they are the best tool for reinforcing common goals and ensuring that ideas are thoroughly debated and communicated. A physical meeting with all the players is one of only a few forums in which the leader is seen as leading the group, as opposed to interacting with individual players. Finally, face-to-face meetings have a human dynamic that is difficult to replace with networked, technologically supported intercourse available through e-mail or groupware.

5.4.1 Virtual Meetings

Before we continue our discussion of what you need to do to ensure that meetings are valuable tools of change, we need to explore the last point more thoroughly. Much has been made of the breathtaking impact that communication technology has had on business and personal relationships, but there is a serious imbalance in the presentation of its benefits. For our purposes, a *virtual meeting* supported by network software (also called groupware) is a meeting even if the sense of time and space is abstract; therefore, we need to understand the dynamics of this new medium as well as we understand conventional meeting dynamics. Groupware is attractive business technology because it lowers meeting costs by reducing or eliminating the number of face-to-face meetings, just as well-written and judiciously circulated memos or well-organized telephone calls do. This technology also seems to create new ways of interacting and to affect interpersonal relationships in ways classical meeting formats cannot. Unfortunately, while groupware both speeds up and improves decision making, its advantages are overshadowed by its shortcomings when technically illiterate managers jump on the bandwagon.

In our rush to take advantage of this new technology, it is easy to forget that many of the natural ingredients of physical human contact are unavailable from synthetic groupware-supported communications. A manager in this new era has to pay more attention to those missing components, of which awareness regulation is the principal one. Mediocre managers deal with awareness needs subconsciously during face-to-face meetings, but even astute leaders can easily neglect awareness needs during virtual meetings. The problem is compounded when change initiatives are underway. Under those circumstances, significant movement can happen only when the actors *as a group* reach comparable activity plateaus often and simultaneously. Time-shifting groupware, however, masks the presence of disjointed, diverging awareness. The extent to which managers learn to use new technologies like groupware to help or hinder the Gestalt Cycle ultimately determines if such technologies can be useful in the particular setting under consideration.

One way to keep a lid on this problem is to replace our soft-headed romance with the endless possibilities of new technology with a focus on technology's practical value and genuine limitations. That requires a healthy dose of skepticism and a knowledge of the history of the particular technology you are considering. Our favorite illustration of romance versus reason is the popular press's infatuation with "virtual" relationships and the emerging transnational cyberculture of the 1990s. The press claimed that cyberspace offered the first opportunity to nurture and maintain fellowship independent of the physical world. They coined this brave new world, *cyberspace.* How innovative! National political structures were being challenged by ideas percolating up through the Internet. How revolutionary! Freedom from tyranny was just around the corner, or so we were told. Lost in the excitement and hype of the moment was that the only thing new about these sorts of relationships and shifts in the power structure was the tool used, not the fact that virtual communities can be created and maintained in an artificial domain.

Virtual communities have existed ever since written language came on the scene. If one of the touchstones of cyberspace is that participants know each other only in an intellectually-constructed, nonphysical world, then the emperors of the Roman Empire were citizens of cyberspace. They relied on a complex communication system to build and maintain relationships with leaders holding power at the far boundaries of the empire. The Roman senators certainly did not meet each one of their adversaries but constructed a relationship with them in Roman cyberspace. George Washington was a citizen of cyberspace, too. Every evening after dinner and brandy, he would retire to his office, boot up his word processor, load his cyberspace software, and continue a lifelong discourse with great philosophers, politicians, generals, and farmers all over the Western world, many of whom he never met and never expected to. In his case his word processor was pen and paper, his cyberspace software the world postal system. It worked quite well for him and still does to this day.

So the real value of technology seldom is that it is new, but that it is an improvement over a previous solution. The essential activity remains the same. The problem set that a new technology addresses remains the same. The future of the automobile serves as a good analogy to the future of cyberspace. In the 1950s, the promise of universal personal transportation was as oversold as cyberspace was in the 1990s. The car was going to liberate everyone, but eventually its real value could not be denied. Sure, you still can use your car to cruise the city streets in search of entertainment, much as you can cruise the Internet in search of a good time, but getting to work downtown while living your other life in the suburbs turns out to be the car's lasting legacy. Groupware's power does not come from letting you wander around cyberspace or allowing you to engage in a political debate with unseen opponents. Its power comes from reducing your cost of establishing and maintaining profitable professional and avocational relationships that would otherwise grow old and cold.

Which returns us to our central premise about groupware. As a new technology, it must represent an improvement on the eternal problem set of meetings. They are the same problems we have tackled since we began talking to each other and no different than the problem set associated with conventional meetings. Groupware's significance will become obvious as long as we do not fall in love with the siren song of the technology itself. With this caution now out of the way, we will return to explaining how to get the most out of a meeting with another person, whether inside or outside cyberspace.

5.4.2 The Chance Encounter

If you are responsible for a project, you must ensure that awareness levels are regularly and vigorously stimulated and that closure and withdrawal are closely followed by additional sensation (see Figure 5.2). If you are a general manager, you may have several significant ongoing change efforts and will be concerned with the awareness of different and sometimes overlapping groups. You, of course, would not have "too many" projects. If you did, you would be violating the principle of unfreezing that indicates that there is a practical limit to the number of changes that can be supported by sufficient and credible disconfirming information.

The only way you are going to be able to keep these initiatives on track is to take every opportunity to boost the awareness both in yourself and in others participating in the change. Regular, written reminders coming out of your office or e-mail are fine, and if that is all you have to work with, that is what you have to rely on. However, nothing is as powerful as personal presence. Every chance encounter you have with subordinates and peers should include a recollection of the project, even if it is a quick recitation. Additionally the Gestalt Cycle indicates that even chance encounters follow the cycle's distinct phases.

Poor person-to-person communication starts with an assumption that the person to whom you are talking has the same foreground figure in focus as you do. That is seldom the case. For example, suppose you start a chance encounter in the hall with, "Hey, Barbara, what's going on with the project?" You should not be surprised if all you get is a blank stare. Barbara has no way of knowing what information you need. She will try to form a figure from her background and hope it matches yours. On the other hand, it is better to start your conversation with a picture of your foreground: "Barbara, I've been reviewing the latest project report and notice that Bill may be having trouble. Tell me what's going on." Barbara's awareness rises rapidly to meet yours. If, on the other hand, you have not done any work on the project in some time, you could start your question with a statement like "Barbara, I haven't looked at the project reports, but I'd like your assessment of where we are." In either case, Barbara does not have to guess what your foreground figure might be. It is not coincidental that reporters usually preface their questions with a statement. It gives their target a better chance of answering the question that follows.

Some managers believe if they telegraph what is on their minds, they will not be able to ask open-ended questions that help uncover what is really going on. They miss our point. It is not that you cannot ask open-ended, undirected questions. It is that *if you want to use your time efficiently* you have to let the other person know the lay of the land. Otherwise, what passes for good, probing give and take is nothing but flailing around for common mental ground.

For example, suppose Girard runs into his occasional acquaintance Marie on the street and asks, "Are you doing anything tomorrow night?" Up to this point, their relationship has been limited to small talk and a quick lunch with mutual friends. Having presence of mind, she will hesitate to answer until she knows why he needs to know. There will be much hemming and hawing, probing, misdirection, and general uneasiness. Maybe Girard likes undefined, tension-filled conversations. However, if he (like most of us) does not, Girard is better off if he starts his question by laying some groundwork. Suppose something she says makes him think she might be looking for something to do. Instead of asking the what-are-you-doing-tomorrow-night, open-ended question, he ought to say something like, "A couple of us from work are going to the ballgame tomorrow night. Do you want come?" That tells Marie everything she needs to know about the question. On the other hand, suppose he has a different impulse. He instead says, "Jenny and I just broke up. Are you doing anything tomorrow night?" Marie has a different picture of Girard's foreground. Whatever is on his mind is unknowable until he tells her. Any question put to her comes bundled with a context. By prefacing his question with what is on his mind, he has a better chance of a clean communication. Of course, Girard himself may not know what is on his mind or may know that he is ambivalent about the situation, in which case he should be prepared for an awkward encounter.

Awkward encounters in business always cost money and should be avoided. Honesty within the company or group pays dividends, so say what is on your mind first and be sure that before closure you have asked about the state of change in your ongoing initiative. We, of course, are excluding activities outside the group, where competitive negotiating suggests that revealing what is on your mind can cost you money.

Besides framing questions so that they convey your foreground figure, all your chance encounters should in some way address the ongoing change project. If you allow brief meetings to drift over the landscape and conclude without exploring the change project, you have lost an opportunity to reinforce the transition state and display your leadership interest in the objective. Even if it is only a parting line like, "Don't forget we have a meeting on Monday to review our work progress on the LMNOP project," you have provided a small boost in the other person's awareness. What is done with that increased awareness is immaterial in the short run, but over the long run such constant, incessant reminders work magic.

5.4.3 The Planned Meeting

There are various types of planned meetings. Without making precise distinctions among the different kinds of meetings, let us observe one interesting truth. Regularly scheduled meetings, like weekly sales or production meetings, seem to be better run than ad hoc meetings used to run special projects. That implies that if we want to make project meetings more productive, they should include the features of regularly planned meetings: a clear objective, time management, an agenda and participant selection, action, and closure.

Those elements address the structure and maintenance of meetings. Other features of good meetings, such as how to conduct them, interpersonal elements, and good agenda-writing techniques, are covered in references provided in the Appendix. We believe that if we insist that every ad hoc meeting have a written objective and agenda and that every project meeting conclude with some written output, project leaders will be able to generate hard, tactical value as well as reinforce the collective awareness necessary to keep change projects on track.

5.4.3.1 Clear Objective

An objective is "something toward which effort is directed." Before meeting participants gather, they should know the precise *expected* outcome of the meeting. The objective statement must be clear and unequivocal. A "better understanding of our problems" might be the purpose of a meeting. In fact, such a statement reflects the role of meetings in adjusting the awareness of participants; however, it is not an appropriate meeting objective. Examples of well-stated objectives include "review of and concurrence with proposed Action Item list," "discussion of current project status reports," "generation of ideas for product XYZ's locking feature," and "review and sign-off on Nonconformance Reports." Such meeting objectives allow better preparation by giving participants an idea of what the meeting is supposed to accomplish and how to prepare for it. Meetings may lead to better understanding of collective problems, but saying a meeting's objective is "better understanding" is a nonstarter. It is a little like your friend telling you to get ready to go out tonight without telling you if you are going to a drag race or the opera. The goal is to have an enjoyable time; the objective is the means by which you plan to do so. Without knowing the objective, you will not know whether to wear comfortable clothes or your finest threads.

5.4.3.2 Time Management

Not only do the most productive meetings begin on time, they also end on time. Ad hoc meetings are a bit more difficult to plan because you are guessing how long closure will take. Regardless of whether you got through your agenda, you must insist that meetings end at the assigned time. If you have to stop short, it is obvious that the meeting's objective was too broad. Under that circumstance, it is always better to withdraw and reflect, even if only for a few hours, so the next meeting can be better focused and controlled. A rule of thumb we follow is that ad hoc meetings held on site should be limited to one and a half hours. Keeping people's awareness on the task at hand for longer than that is difficult when other competing work is just outside the meeting door.

5.4.3.3 Agenda and Participant Selection

Most good meetings are driven by written agendas. The regular weekly production meeting runs off a production form or data sheet. Likewise, project meetings need written agendas, too. Agendas take the meeting objective and break it into manageable, logically progressive activities. Also, writing an agenda makes you decide who needs to be in the meeting, what data are required and who will supply the data, and in what order issues and information will be presented.

5.4.3.4 Action and Closure

As the Gestalt Cycle implies, anytime awareness increases to the degree that a foreground figure emerges, it generates energy for action. Furthermore, closure concerning the meeting gives participants both a sense of accomplishment for the time they spent away from their "regular" work and a clear sense of what has been decided. The action phase of project meetings invariably extracts work commitments from various people. This commitment generates tension in participants that is released only when they complete their commitment and get personal closure, thereby moving the project forward. Of course, if their commitment was only verbal and not psychological, they will leave the meeting with no intention of honoring their commitment and essentially have personal closure the minute they walk out of the meeting. As far as closure is concerned, people get something out of a meeting if they see it as a learning experience. To ensure that participants leave with a feeling of accomplishment, the leader should provide formal closure. A recitation of the commitments and decisions of the group is absolutely required. Setting the time of subsequent meetings helps, too. Merely mentioning that the participants will be notified later of the next meeting is sufficient.

5.5 FREE-FORM ORGANIZATION CHARTS

Another tool for moving change projects forward is the free-form organization (org) chart. The limitations of formal command-and-control org charts (Figure 5.4) are so frustrating that many companies refuse to draw any. Free-form org charts (also called activity-based org charts), like the one shown in Figure 5.5, were developed to overcome the rigidity of formal org charts by limiting their illustration to relationships between people when they are working on a particular process and by using various techniques to reflect the different types of relationships. While not a substitute for conventional

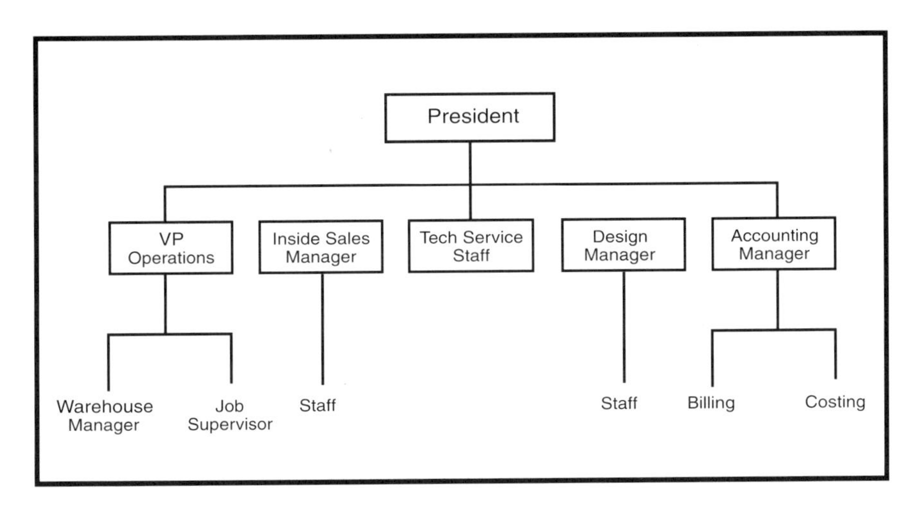

Figure 5.4 Classical organization chart.

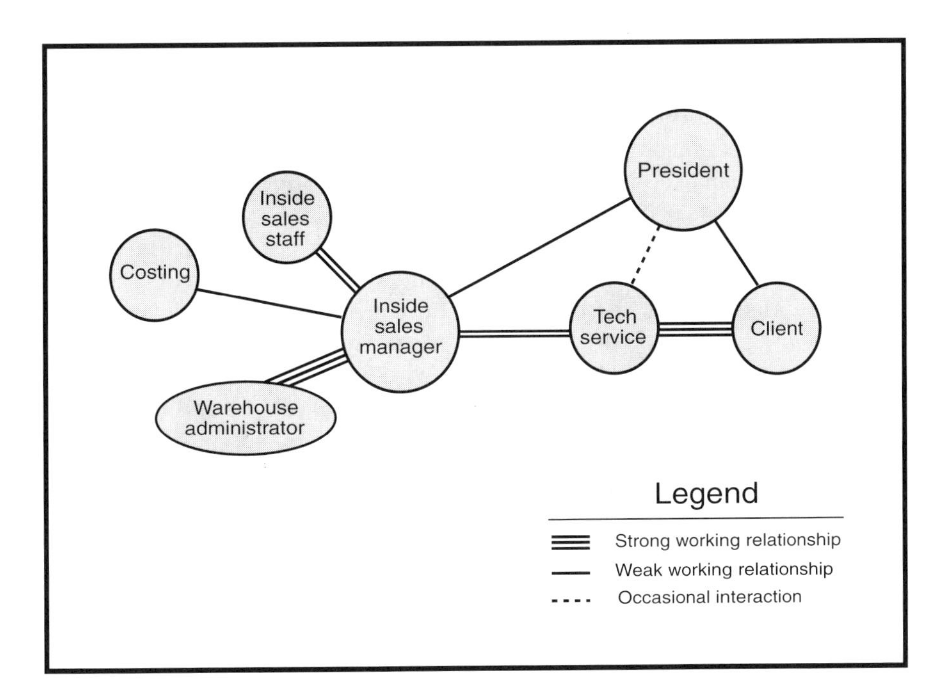

Figure 5.5 Free-form organization chart, showing field service order entry.

org charts, free-form org charts provide relational insights on how work is done.

The recent disreputation of formal org charts is an example of how poor application can lead to rejection of a perfectly good tool. Believing that org charts are commonly misused and therefore useless for illustrating relationships is as silly as believing that keys are useless for opening doors since they are also used to pry open paint can lids. More to the point, the disfavor of org charts says more about the impact of recent management fads that emphasize "open" organization structures than it does about the real limitations of formal org charts.

It is true that the familiar organization chart format rarely reflects true working relationships. However, org charts are supposed to model the performance-responsibility chain, not necessarily working relationships. It is not surprising that they rarely show how actual work is done because they were never intended to do so. Those organizations that depend on org charts—the military comes to mind—know that these charts have a very narrow and specific use. They need the charts because their kind of work requires that each person know, beyond any doubt, who is responsible for his or her personal direction, well-being, and sustenance. In fact, all businesses need to answer those basic questions, but people do not die if the questions are ignored.

Organizations that rely on formal org charts know very well that the informal organization is where most of the real work gets done; org charts seldom show technical relationships. For example, a machinist may report to the shift foreman for a work assignment, as the org chart shows, but he may rely on a senior craftsman for setup assistance and on the president for material performance expertise. The org charts says nothing of those relationships. The org chart may show the quality assurance manager reporting to the president when, in fact, she sees herself working for the plant manager. Few organization charts have arrows showing where the customer enters the picture, although that interface is a critical feature of the company's control systems. These working relationships vary over time and depend on the activity being investigated.

When embarking on a change initiative, you should model the informal organization using a free-form org chart. It will help you see who has a stake in the changes and whose participation is essential. Suppose you are going to reduce the number of steps and hold points in your field-service order entry system. While you trace the path of a typical service order from customer call to work order issue, you will come up with a list of people who play a role in the process. At the center of your org chart, draw a circle with the name of the predominate person involved in the process. As you draw other circles representing other actors, you may decide that not one but several people occupy the center of the activity; therefore, you might place several near the center. As for representing the relationship between people, you could signify a strong working relationship by two solid lines, a weaker relationship with one line, and an occasional interaction by a dashed line. Some people like to use distance from the center to represent influence on the process, but that can make drawing the chart more difficult.

When you finish your free form, you will see relationships that are impossible or inappropriate to show on the formal org chart. You will have a better idea at what point and with whom the customer interacts. Realizing the nature of the organization behind the organization is essential for identifying who can influence the direction of the change you want to implement.

Refer back to the free-form org chart in Figure 5.5. That example is based on an actual case. Several relationships are evident on that org chart that were transparent on the formal org chart in Figure 5.4. First, the free-form chart applied only to the field-service order entry process, not the entire organization or all the processes. Second, the president's link to the technical services people was a weak one even though the formal organization chart shows them reporting directly to the president. For the work under consideration, the technical service personnel had a stronger relationship with the inside sales manager than they did with the president, but only as far as the order entry process was concerned. Third, the client rarely, if ever, interacted with inside sales, which in this company served more as internal administrative support than a tool for the customer's use. Next, the inside sales staff interacted almost exclusively with their supervisor, not with the outside sales staff or the customer. Last, there was a strong relationship between the warehouse administrator and the inside sales manager. The administrator did not even show on the formal organization chart. During interviews, it became apparent that the inside sales manager developed an informal relationship with the warehouse so that she could get vital information on scheduling and spare part stock levels unavailable through the formal organization. The formal organization chart reflected none of these working relationships, but then it was never intended to do so.

Another use of free-form org charts is for general diagnosis of the current state of work relationships. At the beginning of a major change initiative involving the entire company, the project leader should sit with the major players individually and draw a free-form org chart with the interviewee at the center of the chart. Instead of focusing on a specific function, ask each person to think about his or her working relationships in general. Say you have six key people. You will wind up with six free forms. If there are significant differences between the charts because certain people see strong or weak relationships where their counterparts see the opposite, then you have radically different perceptions of how work is accomplished. If your change project depends on those relationships for success, you will need to address the misperceptions directly before you can proceed with the transition phase of your change project.

5.6 FORCE-FIELD ANALYSIS

Another popular tool used to illustrate the dynamics of change is the force field. A simple tool, the force field shows the forces aligned for and those opposed to change. That information is invaluable for maintaining change momentum since it allows you to determine what opposing forces must be removed, as well as which supporting forces might be helpful. Removing opposition turns out to be more critical than enhancing supporting forces. This is similar to salespeople knowing that they must identify, address, and isolate all objections before they can close a sale. To change a company, you have to sell an idea first.

Figure 5.6 is a free-form diagram for the change initiative associated with the field-service order entry project cited earlier. The line down the center of the diagram represents the present state that must be metaphorically moved to the right. The items on the left are the forces for change, those on the right oppose change. The size or number of the arrows associated with a force indicates its strength relative to the other forces. In most instances, you will come up with more opposing than supporting forces.

In the real-life example in Figure 5.6, we identified three opposing forces that had to be removed if we were to have any chance of seeing the project to its conclusion. First, the company was scheduled to move from

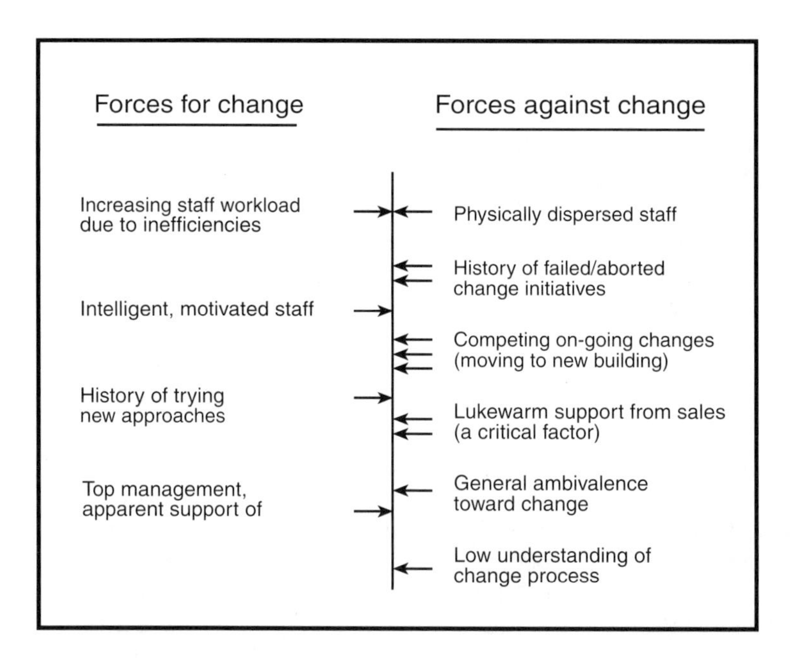

Figure 5.6 Force-field diagram.

dispersed operations around the city to one central location in six months. We questioned whether it could endure such a disruptive change and still remain focused on this relatively small but important project. The president decided that they would merely suspend work on the field-service system during the most intense activity period associated with the move. Second, the company had a history of starting change initiatives and then letting them dissipate. None had been successful. We were concerned that that history of failure would color people's enthusiasm toward this latest change project. This was a distributed opposing force, not centered in an individual, and could only be overcome by deed. Third, during the early part of the project, it became evident that the sales department representative—a critical actor in the change effort—had little interest in the change. She was uncomfortable with her central role, had little training in or understanding of systems analysis, and perceived few rewards for her efforts. The president was aware of this but saw the project as one that could also help this staff person overcome her shortcomings.

Not surprisingly, this project failed, too. The move to a new location was inadequately planned, exhausted the staff, and, because of cost overruns, left the company with less financial breathing room. The move also introduced tremendous amounts of disconfirming information about all phases of the operation, not just order entry, and thereby overwhelmed management. Unfortunately, the avalanche of bad news disoriented rather than focused the president and his staff. The planned change project gave way to ad hoc, unplanned firefighting.

That the force-field diagram "predicted" the outcome is immaterial. What is important is that those involved in the change effort identified the opposing forces accurately and addressed them as best they could. Without the benefit of generating the force field, discussing the significance of the forces aligned against the project, and addressing these forces prior to the undertaking, the project had *no* chance of success going in. During the formal closure phase of the project (which incidentally never happened), a review of the opposing forces and their role in the failure of the program would have reinforced the idea that the probability for success was directly related to those known forces.

5.7 MEASURING CHANGE

Just because you cannot measure something does not mean it does not exist. Love, hate, beauty, quality, goodness, and evil all exist even though we cannot measure them objectively. In the circumscribed world of business and profit, however, you must believe just the opposite: if you cannot measure it, it does not exist. That fact, and the frustration that goes along with it, plays havoc on rational people. In a 1995 headline article in *CFO* magazine, financial leaders of companies such as General Electric, IBM, AT&T, and Pepsico relied on hyperbole to express the financial success of reengineering projects. Such was their total detachment from reality that some were quoted as saying that most of the benefits of reengineering were unmeasurable, but nonetheless they believed reengineering was a resounding success. It is a sad day when the Kings of Quantitative Rationalism run out of creative ways to illustrate the value of a significant and costly undertaking like reengineering.

Let us restate the maxim to "If you cannot measure it, do not bother with it." If you believe that programs like reengineering defy measurement, do not undertake them. If someone else in another company can measure a program's benefit, it is not a poor reflection on your company that you cannot. It indicates that either you lack the understanding of the program's costs and benefits and therefore should not undertake it until those costs and benefits become more evident, or if you do understand it, you should not undertake it because it holds insufficient benefits for you. On the other hand, maybe those competitors or friends undertaking the latest fad management program are in such poor condition compared to you that they can get value from it even if you cannot. Or maybe they are in such poor condition that they have no idea that their profitability model is bogus. In any case, your mother was right. Just because your friends are jumping off the Brooklyn Bridge, why should you?

Of the different skills required to get a program through the implementation-transition phase, figuring out what to measure and how to value it requires the most from your creative and quantitative skills. You should be concerned with these two issues:

- What the project will cost and what monetary benefits it will generate;
- What you will measure to show that the project had its intended effect.

The second point differs from the first in that you may have to develop a surrogate measurement that indicates the project's outcome rather than measure its outcome directly.

We will use the example of improving the order entry cycle time to show how you might approach a real-world problem. The costs of undertaking the program are mostly opportunity costs, that is, your people's time. Instead of working on this project, they could be working on another, so if they work on this particular project they forgo the opportunity to work on something else. Their time is not free. Before you start the project, you must estimate how much time will be required for the staff to do the analysis, write the new procedures, test them, train employees, and implement the new work rules. A good rule of thumb, if you do not have a track record for estimating the work load for these types of projects, is to try to think of all the steps involved, estimate the time required, then double it. Now you have an estimate of how much time and money a project will cost.

Ignoring opportunity costs is the first place people go wrong. Some will say that the project is cost-free since the people that would be working on the project are already on staff. That reasoning concludes that since salaries are, within reason, a fixed cost, the project is free. Nothing in life is free. Even if the staff is salaried, all incremental work they perform costs the company real money. Opportunity costs are costs associated with doing one thing rather than something else, and they are real. The fact that you cannot imagine what they would be doing if this project did not come along results from limited imagination, not from limited work opportunities. If you indeed had no inventory of profitable project work, you soon would realize that you are overstaffed.

Back to our example of reducing the order entry cycle time. Suppose you figure it will cost $25,000 to get all the way through the program. Now you have to quantify the benefit. Some will say that it is obvious from inspection that reducing order entry cycle time is a profitable venture and spending time on quantifying the benefit is a waste of time. Not so. Look at the ways you could *increase* costs when you decrease order entry cycle time:

- *Increasing the head count.* Suppose you go ahead with the problem analysis without worrying about quantifying its benefit. Suppose you determine that the only way you are going to decrease order entry cycle time is by hiring more order entry help because "everyone is loaded." You have just added structural costs without showing how it will be paid for, much less how it will make you money. But you have faith that any reduction in cycle time by definition makes money.
- *Increasing efficiency.* Some people might balk at adding overhead but nonetheless claim that in their world increasing efficiency by definition makes money. Suppose your analysis shows that you can shorten the cycle time by streamlining the number of order entry review steps, thus freeing staff from that function. You do not lower costs yet because that freed-up staff time is reallocated to other activities, ones that you are convinced need to be addressed but for which you never had time. Yet you have faith that any reduction in cycle time through increased efficiency in any part of the company has to make money.

In both cases, cycle time was reduced and costs certainly went up, at least by an amount equal to the time spent studying the problem and instituting the fix. But if you had to show a profit from those changes, you would have to show how shorter order entry cycles flow to the bottom line. You would eventually wander over to take a look at the average length of the job queue sitting in the field-service in box. If it is longer than a certain critical length, field service can never take advantage of faster order cycle times coming out of the office. The improved cycle time in order entry will be absorbed in a longer average queue time in the field service backlog. (Readers with an industrial engineering or operations background will recognize this classical systems dilemma; however, it is surprising how many who have appropriate training in this area miss the analogies between factory queuing problems and overhead queuing problems.)

This example leads us to the basic, twofold problem with failing to generate a hypothesis about how a project will make money. First, you will miss important ramifications because you have not been forced to think the problem through. Second, you will not be able to develop a robust measurement strategy capable of showing that costs were under control and that the promised profits were delivered.

To continue our example, before you start your study you have two possible hypothetical sources of profits from its outcome, regardless of what action items the study generates:

- *Reducing the head count.* You can hypothesize that your study will free up staff time while overall cycle time remains the same. If so, you then know the only way you are going to make money from this subsystem improvement is that you must put the excess staff to work elsewhere. Either that or they will have to be eliminated from the income statement.
- *Increasing sales.* If, on the other hand, you could believe that reduced order entry cycle time will cascade into the field-service staff and result in a reduction in overall cycle time. In that case, you have to translate that hypothesis into a plan for utilizing the increased capacity by increasing sales. When sales increase concurrently with a reduction in order entry cycle time, the marginal profit on the additional sales is better than the current unit profit because you do not have to add staff to take care of the increased sales. However, if sales do not increase and you decrease cycle times (thereby increasing capacity), you have no return for your improvement project.

Either way, before you decide to study how to reduce order entry cycle times (or any other process improvement study), you have to have a firm idea what you expect to find and which related activities will be affected. Of course, your hypothesis could be dead wrong. (We will show in a later chapter why it most likely will be wrong!) But like the design of any experiment, the value of a hypothesis is not whether it is right or wrong, but that if you do not have one you can reinterpret the outcome of the experiment any way you wish. Without establishing this intellectual anchor before any change is undertaken or any test applied, the conclusions you generate are illegitimate. That is where individuals with science or engineering backgrounds are supposed to have a competitive advantage over managers coming from other disciplines. Unfortunately, many bright technologists seem to lose their minds when they take up management duties.

The role of the hypothesis in science is translated into a going-in profit model in business. Without that concept, you may have a successful program—that is, reducing cycle time—but receive no benefit from it. The discipline required to generate hard numbers also forces the planning staff to answer hard questions before the project is underway. If your going-in hypothesis is that a project will result in a staff reduction, certain consequences are implied. If the hypothesis is that it will result in additional capacity, different consequences are implied. The accuracy of the calculation is not as important as your plan to capitalize on the change.

A favorite example of spending money on a pet project without showing a return is the ever popular customer questionnaire. Some of today's faddish quality programs imply that to be a world-class company you have to have a program of regular customer surveys. Quality departments regularly insist that they need surveys because everyone has one. Besides, customer surveys are a "cost of doing business" that cannot be justified on a current income basis. Not so. This is how the cost-benefit argument could be structured: First, someone calculates how much the survey will cost, then doubles it. Next, we have to realize that the benefit from any data program, whether it be a customer survey or a plant data system, is not the data itself but what you do with the data. In the example of a customer survey, you have to ask yourself what the data will allow you to do or what costs they will help you avoid. If all you do is generate numbers indicating how your customers feel about you with no working hypothesis of how that information will be used, you have just spent money on a survey without any resulting benefit. If, on the other hand, you design the survey to help you decide your next product improvement or prioritize your work, then the cost of the survey is rolled into the cost of the anticipated actions you will take based on the information you gather. The root problem with activities justified as a cost of doing business is that no one takes the time to understand how the company actually makes money with those activities. Without going through the exercise, it is easy to add the cost and forget to generate value from it.

While it is easy for us to tell stories of how project costs can be rationalized, it is harder for you to apply them to your particular situation. As a manager, you need a few basic back-of-the-envelope steps to apply to any proposed project. If you work for a larger company, you may already have a standard format for pencil-whipping a project into a shape presentable to your investment board. Nevertheless, what separates the good project from a wish list is the character of the analysis, not the neatness of the presentation. So whatever project you run into and no matter its size, apply the following measurement elements:

- *Generate a written estimate of costs.* Remember, nothing is free. Also remember to double your cost estimate. Even if you overestimate the cost, any improvement project worth doing can support a doubling of its costs without jeopardizing its success. If you are presenting the numbers for someone else's approval, do not double the number, but show in your back-up that the costs may double without jeopardizing the project.
- *Establish how the improvement will generate profit.* Do not forget that if you squeeze costs out of a system, those costs have to be physically removed. Otherwise, it is just funny money. If you increase capacity, remember that you have to have a plan for loading that freed-up capacity. There is little to be gained by cutting cycle times in half (thereby doubling plant capacity) unless you can load the resulting idle capacity or sell it to someone who can. Otherwise, it is just excess capacity.
- *Determine how those costs will be tracked.* How will you know how much a project costs you if you do not have a way of tracking expenses? If you currently do not have a project-centered cost system, simple daily time sheets showing how much time someone worked on the project will be enough. You are not trying to split the atom, you are just interested in measuring cost within a reasonable error.
- *Decide which project activity measurement reflects progress.* Many corporatewide and departmental improvement projects involve increasing the efficiency of work, especially overhead functions. While an investment in plant and equipment generates obvious progress indicators (e.g., "The pad has been poured, the machine set, and now we are testing"), measuring progress on projects aimed at increasing efficiency is a challenge. Whatever the activity, you have to decide how you will be able to measure the impact of the project on the work activity. Many measurement schemes used to track improvement programs show progress even though most people involved agree that they have generated a lot of heat but little light. How many teams you have established, how many procedures you have written, or how many plants you have toured is no indication of project progress or completion. If your objective is reducing cycle time, then you have to measure cycle time to have any chance of knowing whether your project is making progress toward its objective. If your project is decreasing rework, then you have to have a good system to measure rework.

The fact that you have trained 83% of shop personnel in statistical process control is irrelevant.

- *Ensure that adequate baseline information is available.* Many projects begin without the benefit of knowing baseline performance. Take time to generate good baseline data before you start your journey. For example, if you are attempting to reduce scrap, how good is your current baseline information? Is your current scrap report accurate? Is it sufficiently precise to measure any improvement you might make? If the information is solely anecdotal, then you have to spend time setting up your information sources before you proceed. You might think that will hold your start time back, but if you have poor data, you probably have other opposing forces on the force-field diagram that need attention before launch time anyway.
- *Ensure that measurement methodology is in place and reportable.* If your information system was in place and generating good baseline information before the project started, ensuring that you will continue getting relevant data is not too difficult. However, if you had to set up new measurement methodologies, remember that you have to schedule time for training people and testing measurement reliability before you turn your attention to other details.
- *Issue a financial analysis at the end of the project.* If you are the owner of a company and run the project yourself, maybe you do not need to document the project's success or failure. You are intimately involved throughout the project and know what the numbers say. But if you are spending someone else's money, you should write a summary report showing the results of the program. That will not take much time if you have kept up with all the other program measurement activities. If you have not, you can claim the project was successful, but you cannot prove it.

5.8 GENERATING OPPORTUNITY

That completes our recitation of some of the tools you will need to organize a change project. Our assumption up to this point has been that you have already selected the appropriate change project. But where do the ideas and opportunities for change come from?

In Chapter 1, we said that peculiar to the role of the manager is the responsibility to operate on the system. That requires that at times you as a manager have to go outside the system to be able to work on it creatively. One key can be found in the lesson of the Gestalt Cycle. In the rough parlance of Gestaltists, you have to practice suppressing your natural habit of forming figures from background noise if you want to free yourself from automatic thinking. By doing that, you will see more change opportunities than you would be able to otherwise.

How do you do that? First, using the Gestalt Cycle is a conscious activity, not the result of some mysterious Eastern enlightenment. You have to practice self-awareness. Try a little experiment. The next time you tour your domain—say, your department or the shop—do so without an agenda. That is to say, go out to the shop or tour your department with the agenda of not having an agenda. Do a lot of listening. As soon as you see yourself forming a familiar foreground picture, drop it. Instead, pay attention to details you have never paid attention to before. You will find yourself noticing things you have grown used to, such as peeling wallpaper, drafty rooms, or poor lighting. Maybe you will sense tension and disillusionment. Perhaps you will be impressed by the intensity and the focus. Whatever the case, you will see artifacts you have not seen before. More important, do not go back to your station deciding to address some issue you have just dug up. Let it go. This will frustrate you beyond belief because you have not allowed yourself to form foreground figures nor to get closure. Later, when you are working on previously identified problems and applying the principles of planned change, something left over from your unfocused wanderings will float to the surface and have more meaning than if you had tried to force an issue to the surface at the time you observed the artifact.

Several years ago a management fad called *management by wandering around* (MBWA) became popular through the book *In Search of Excellence.* Scads of managers flocked to the shop floor or to ground zero of their departments to try their hand at unfocused, undirected management. But like all passing business fads, the technique was oversold and poorly applied. It is extremely difficult work and requires strict discipline on the part of managers to refuse to draw inferences from what they see. Most of the new MBWA converts did nothing of the sort. They went out to the floor with an agenda. They allowed themselves to get caught up in a firefight. While they might have felt refreshed when they got back to their stations, they missed the point. In the end, the fad was blamed for all sorts of company disruptions, which were inevitable because management staff were not sufficiently unfrozen and ready to change their style. There was no overwhelming threat to energize them. They failed to prepare themselves with responsible training, and they mistook their goal ("We need to improve our management style") for their objective. Like all fads, MBWA started with good intentions but was perverted by managers who lacked the fundamental skills of manageering.

Chapter 6

Controlling Change

At this point, you have gained an appreciation of why it is so hard to initiate, sustain, and complete change. You even have a handle on how to identify profitable change objectives and test whether you are making headway. But opportunities for change present themselves almost every day. You cannot react to low-level disconfirming information in an ad hoc way. You have to have a rational way to process those opportunities. Otherwise, disconfirming information boiling up from below either will go unnoticed or will create a management team with a firefighter mentality. That is not a healthy way to run a company. You need a system that anticipates potential sources of disconfirming information, categorizes them, and has planned responses depending on your own criteria. But how do you do that?

6.1 CONTROLLED VERSUS UNCONTROLLED CHANGES

To understand the variety of changes companies struggle with every day, the first thing you need to do is group change opportunities into two domains, controlled and uncontrolled. A *controlled change* is one that is guided by well-defined and (one would hope) written procedures. The fruits of our labor on controlled changes are newly defined or revised activities that are established and maintained as part of the accepted standard operating procedures of the company. The resulting procedures are controlled by planned review process and become part of our competitive advantage. An *uncontrolled change* is one that happens without the benefit of systemic awareness. That is, the resulting procedures are seldom reviewed by anyone other than the instigator, and few if any records are made of the newly designed activity. (Some observers refer to uncontrolled change as *naturally occurring change*.)

These two types of change can be envisioned as loads on a balance scale (Figure 6.1). When a good balance between controlled and uncontrolled change is struck, the company's change processes are themselves under control. Whenever either side of the balance is overloaded, the company's change processes are out of control. In the simplified example in Figure 6.1, a fictitious company has defined order entry, field performance feedback, plant scheduling, and new product design as activity domains for which controlled change will be enforced. Likewise, it has decided that purchase orders, production drawings, and machine setup are activities that are best left outside the controlled domain. This example is near-reality for companies producing low-grade, loose-tolerance commodity machined parts. Order entry is controlled by use of a preprinted order entry form; plant scheduling is posted on a whiteboard; all customer complaints are logged, reviewed, and resolved; and new product design is reserved for the chief engineer. On the other hand, they rely on verbal purchase orders to get their needed materials, on penciled changes applied to production drawings to take care of customer change orders or production floor problems, and machine setup by 20-year veteran craftsmen to ensure an adequate level of quality.

While this may not be the split in your company, every company establishes its own division between controlled and uncontrolled domains, depending on its own understanding of the costs and benefits of over- and undercontrol. In practice, what is considered the controlled domain is usually well defined and widely acknowledged, while everything else is by default dumped into the uncontrolled domain. The wide differences between how competitive companies are organized can be traced to their individual decisions about which areas need control and which are better left alone to "empowered employees."

Suppose the company in Figure 6.1 decides that it is spending too much money on maintaining its field performance system and getting too little in return. Every week they sit in a meeting and review the failure reports, and nothing seems to get done. Each week the same failures are reported, once in a while a new one is posted, and eventually the older ones die of neglect. The general manager decides to scrap the field failure reporting system and tells the service personnel to take their problems directly to the plant manager to get them resolved. He figures that those people working together, empowered to make "necessary changes," will whip this pup into shape. He has made a conscious decision to remove the systematic field performance reporting and resolution from the controlled domain. If the division between controlled changes and uncontrolled changes is near balance, a significant change like this will upset the company's control systems and lead to an undercontrol situation [see Figure 6.1(b)].

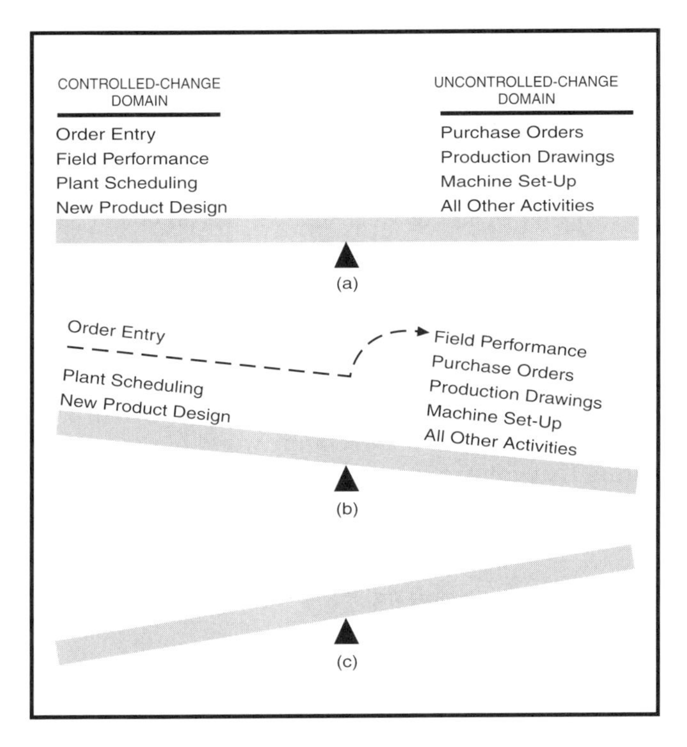

Figure 6.1 The balance scale of change: (a) balanced, (b) under-controlled, and (c) over-controlled.

On the other hand, suppose this same general manager decides to pull machine setup into the controlled domain. He has studied the field service reports and decides that too much leeway in machine setup has led to an unacceptably wide variation in product attribute. His solution is to get control on how machine operators set up their equipment. Analogous to the decision to remove controls over a significant element of company operations, putting controls over heretofore uncontrolled elements can cause systemic constipation and overcontrol [see Figure 6.1(c)]. Unless well planned and executed, with appropriate attention to increased cost of control, plant production could easily slow to a crawl, thereby discrediting the change initiative and putting the company in a worse situation than before the change was attempted.

Neither decision by itself was necessarily wrong. They were wrong in the sense that the company was in balance in light of the control assets it had. If failure trends and disconfirming information had indicated that the company was undercontrolled, management obviously had to apply control in the appropriate areas and vice versa. If the company is in control, then adding or removing elements from the control system, without changes elsewhere, will lead to an out-of-control situation.

This example assumes that change opportunities present themselves one at a time and that management knows how far in or out of balance its control system is. Neither assumption is valid in the real world. The key to overcoming the reality of change is to ensure that all change opportunities are at least put in the right domain. If you put too many change opportunities into the controlled change arena, then you certainly will overload management and stifle front-line personnel. If you relegate too many change opportunities to the uncontrolled arena, you will be overwhelmed by suboptimum solutions and long-term dysfunction. While an overbalance in either direction leads to declining profitability, most American management teams today tend to put their faith in the liberating world of uncontrolled change, thereby ensuring that companies operate in undercontrolled states most of the time.

6.2 TWO REAL-LIFE EXAMPLES

To understand the consequence of inappropriate reliance on uncontrolled change, let us look at two real-life stories of lost opportunity. In the first story, Vern is responsible for moving semifinished product from station to station for a company that designs and builds large processing vessels, the kind that have to be transported horizontally on flatbed trucks, by train, or on barges. One day, Vern gets a move ticket telling him to move a newly built shell from the welding shop to the head joining shop. Vern takes his crew over to the welding shop. They soon realize that the fabrication drawing did not include lift eyes. This has been happening quite a bit lately. The first few times they ran into this problem, the production foreman told Vern to have a welding crew tack on some lift eyes. So Vern does the same thing this time. Two months later, the completed tower rips the bottom off a highway underpass when a lift eye does not quite clear.

Vern's change process was uncontrolled. Vern found that the method for moving material was consistently unreliable. It had to be changed for him to do his job more efficiently. To solve the problem, he initiated an unplanned and therefore uncontrolled change. He took responsibility for attaching lift eyes when they were not available.

That type of uncontrolled change occurs all the time in business. People are continuously subjected to internal and external disconfirming information that demands attention and resolution. Certain types of problems will not go away unless they are addressed: billing problems, payment problems, inspection problems, equipment problems, sales problems, every type of problem that has to be solved for the company to move forward, to complete a sale. Uncontrolled, naturally occurring changes are those that people make "naturally." Over time, people learn which problems are considered their domain and which must be left to others. Seldom are the boundaries of those domains described formally because there is no formal, recognized process for reviewing disconfirming information, generating creative alternatives, or communicating proposed solutions to people who work at the boundaries of decisions. Uncontrolled change domains can exist totally within one person's job or can involve people in several departments. The missing ingredient is that formal (i.e., written) change procedures are not engaged when they are appropriate to the problem at hand.

In Vern's case, he was given the responsibility for doing a job. He did it the best he knew how with the best intentions. In fact, he had faced the problem of lift eyes so often that he changed his work routine. He now was ordering out welding work. No one seemed to mind. In fact, the foreman was happy that Vern was showing initiative. The weakness in the system was that Vern had no way of communicating the need for a systemic change, no tools to direct the change itself, nor any method for analyzing the potential effects of his solution. When the vessel got wedged beneath the overpass, Vern was the one left holding the bag. Maybe the final inspection person caught some heat, too. However, it was not Vern's job to care and feed the system that ensured controlled change kicks in at the appropriate time, nor was it the foreman's. It was job of the general manager, the person who was supposed to manage the boundaries between functions and build the company's business systems. Unfortunately, many general managers ignore that dimension of their jobs because they do not have a clue how to direct systematic change, a fundamental management skill.

6.2.1 It Is Not Always Vern

Management always likes to point to Vern as an example of how things can get screwed up, not realizing that Vern seldom is the problem. Management is solely responsible for Vern's predicament. Another true-life example will show that management regularly reacts to disconfirming information with uncontrolled change, too, in situations that absolutely demand controlled change.

Irving was recently recruited by the chairman of a Fortune 500 company to head one of its manufacturing subsidiaries. About three months into his tenure, Irving got a call from the chairman. "Bud," the chairman said, forgetting Irving's name. "I've just been talking to Lynn (another division manager) at the Triple D division. They've come up with a great idea to save money on their manifolds. They've reduced the number of weld passes and really increased throughput." Irving realized he had to get with the program. His disconfirming information came from an external source and was not so easily ignored. The information was loud and clear: "Irving, you are not doing your job as well as your peers. Show some initiative." The disconfirmation was all the more pointed by the chairman having already forgotten his name. Irving was the kind of guy who does not need as much psychological safety as some people. He had gotten his job by being hard-driving, taking risks, and generally bulling through any problem that got in his way. By God, he was going to make some changes around here.

Irving thundered down to the production line, talked to the welding foreman, and told him to make the changes that the Triple D division had made. Sure enough, after they made the change, overall throughput soared. That was where the production bottleneck was. Irving was more than eager to share his triumph with the chairman, especially because he was able to make the change so quickly. He might not have seen the opportunity as soon as his peers had, but once on the job he got it done immediately!

This story has two possible endings (telling the actual result would spoil it). In the first ending, manifolds begin to fail at a higher rate than before. In fact, the trend is not recognized for a year or so. It seems they have developed a nasty habit of blowing up in the field. Of course, it takes at least a year for anyone in the company to accept the externally generated disconfirmation ("Your manifolds are failing..."); even then it only gets the attention of the sales organization. After all, as far as the engineers know, nothing has changed. Why should their manifolds be failing? The last time they thought they had a problem, they found that those crazy customers were misusing them. Everyone seems satisfied with the answer, except the customer. Time marches on, and Irving's manifold sales start to drop. "Just a bad market," everyone says. That is true, the market is shrinking. Unfortunately for Irving, the excuse of a declining market masks the effects of the uncontrolled change Irving had made a year before. When the market does revive two years later, Irving's manifolds do not. Triple D's, however, do. Eventually, Irving's division is sold off. One of the first things the new owners do is to increase the number of weld passes.

The problem was that, while the Triple D division was building the same manifolds that Irving's division was, Triple D was selling them into a less demanding market. Triple D could lower the shock-design safety factor without any negative consequences. Their manifolds never saw service anywhere near their nameplate performance. But Irving's division sold manifolds into markets that operated them near their operating limits. Irving's replacement called for the engineers to perform a full-blown failure analysis. They quickly discovered the problem but could not explain how such a design change could have been implemented without verification and validation reviews. No one left over from the previous regime could remember how the change was made since there were no records. The lesson remained unlearned.

The second possible ending to this story starts back on the same day Irving directs his uncontrolled change. Robbie the engineer is wandering the shop floor the same afternoon Irving tells the welding foreman to change the design. But Robbie does not witness the directive. Instead, she notices that the pipe fitters are not preparing the material as they usually do. She does a little digging. When she gets to the welding foreman, she finds out that Irving ordered the change. She tries to get the foreman to rescind the change since she realizes that the manifold's capacity to withstand shock load has been compromised. The foreman says there is no way he is going to tell Irving. So Robbie decides to call Irving directly and let him know the consequences of his uncontrolled change. Irving rescinds his change, but for two years Robbie's career goes nowhere.

Irving acted with the best intentions on external disconfirming information by forcing through an uncontrolled change, the same as Vern. The difference is that Irving should have known better. He knew that the type of change he wanted to make is typically addressed by a systematic and planned change process. Furthermore, anyone familiar with manufacturing knows that production facilities are organized to anticipate design changes and that a well-run shop has systems in place that exclusively handle the type of change Irving had unilaterally imposed. Vern was only doing his job; Irving was derelict in his.

6.2.2 Going Outside the System

Irving's reaction to disconfirming information is typical of many managers who have expensive control systems already in place. They ignore them. They think that the systems are too burdensome for people of action like themselves. Besides they do not have time to learn them. They have things to do. Heaven help the employee who orders a pencil without going through channels, but if the boss wants to buy a new, $10-million building, he need not follow protocol. The boss is empowered, can get things done. "The guy who had this job before me is not here, so he must have been doing things wrong. I'm going to make some changes." The reason policy and procedure manuals fall into disuse is not because workers ignore them but because management does. Because they have the raw power to affect uncontrolled change with far-reaching effects, they do so to the detriment of the whole company.

6.2.3 The Long-Term Effect of Relying on Unplanned, Natural Changes

Over time, companies like Irving's and Vern's go bankrupt in spite of the best intentions of their people. Figure 6.2 shows the relative volume of transactions (T) and uncontrolled-change volume (C) that a typical company sees during the three stages of a business cycle. During the first stage, the company enjoys a certain level of activity. We will assume that it is profitable. Behind the scenes is a set of problems, represented by the companion bar labeled C, that are held in check by uncontrolled, naturally occurring changes put into place by the people who do the real work at Vern's company. Those problems are solved episodically. Some solutions are better than others, but the company manages to get by. Unfortunately, the inefficiency resulting from that problem-solving mode is hidden by the continuing profitability of the company. Lucky for them, their competitors' approach to running their own businesses is about the same. As problems come up, management empowers lower levels to work out those problems as they see fit.

During the second stage of the business cycle, company sales increase rapidly, as does transaction volume. Maybe the market is good; maybe the company develops a competitive product advantage. In any case, debt is increased so that production facilities can be added. More people are hired to handle the increasing number of problems inherent in growth. Profitability is maintained. There is an uneasy feeling that the marginal profit on the increased sales should be higher than it is, but, for the most part, people are satisfied with performance. Behind the financial numbers is a policy of solving problems by relying on uncontrolled, naturally occurring changes. Management rarely gets involved in problem solving. After all, they believe they hire "good" people to work out problems on their own. The volume of

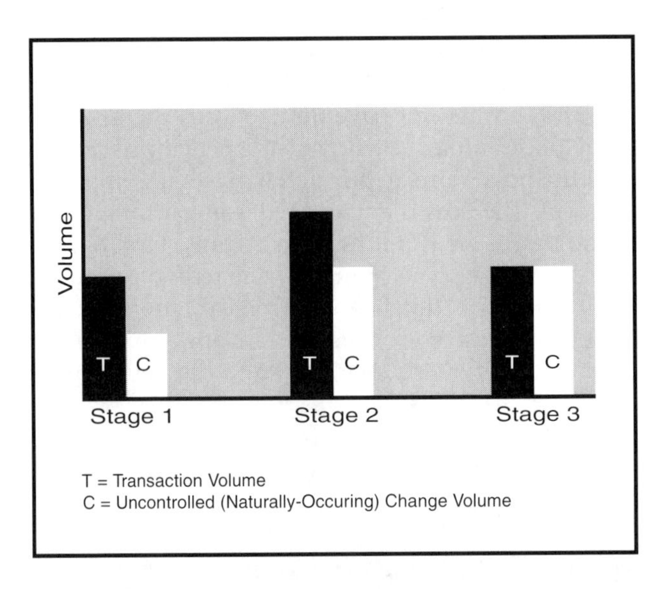

Figure 6.2 Typical business cycle (effect of uncontrolled change).

problems kept in check by naturally occurring changes increase as transaction volume increases. No one notices anything wrong.

The third stage of the business cycle is the downturn. Eventually, all businesses experience a downturn; as they do, transaction volume falls. But naturally occurring changes only hold the root cause of the underlying problem in check; they do not permanently resolve them. The solutions usually are so frail that any change in the environment will tear them to shreds. So the number of problems do not fall as fast as sales fall. In fact, the volume of problems may actually grow. As the third stage in Figure 6.2 dramatically indicates, neglected problems can overwhelm the company, driving down the profit to critical levels. Management rationalizes the company's failure by claiming that the drop in sales was so overwhelming that no one could have stopped the inevitable failure. But they conveniently omit the fact that some players in their market are still in the game. If they do consider that fact, they assign survival to luck when, in fact, the ones who do survive are the ones who used planned, controlled change effectively to solve their systemic problems when they had the luxury of a good market.

6.3 KEEPERS OF THE SYSTEM

Policy is the device managers are supposed to use to decide which problems should be solved through systematic review and controlled change versus which ones should be left for empowered employees. A management prerogative, policy guides which activities are addressed by the systemic procedures of the company and which are reserved for craft and skilled workers. That is why policy is the friction point between all disputes between business owners and organized labor. Policy says not only where but how the assets of the company will be applied.

With benevolent company leadership and employee empowerment being all the rage, traditional management seems to be on the defensive, always apologizing for exercising its prerogatives and running for cover under the latest business fad. But regardless of all the apologists and the real and legitimate influence of organized labor, managers continue to have a crucial responsibility to their company. It cannot be delegated, ignored, or wished away. Managers are the keepers of the system. They have to ensure that the company's control systems are well developed, well fed, and well understood because everyone relies on the system to answer critical questions of survival and growth.

What we call *the system* is shorthand for the collection of actions and reactions each company goes through when facing a problem. The problems can be as simple as getting the product out the door the afternoon your foreman is out sick or as complex as answering a price cut by a competitor. It can be as visible as the obvious division of labor between the front and back offices or as veiled as the corporate culture. Nonetheless, the system exists, and management exists to run it.

6.4 UNCONTROLLED CHANGE: THE ROOT OF ALL FAILURE

In the 1970s, the word *control* took on an almost evil, exploitative connotation. Business gurus invented fads popularizing the notion that systemic chaos was preferable to control. Faced with the increasing chaos of the marketplace and perceived rebellion of the era, management fads embraced the idea that businesses were by nature unmanageable and that the role of management, therefore, was to emphasize flexibility by removing controls. Those pro-chaos fads rationalized their premise by using logic that ran something like this: American companies are in trouble. The common modus operandi consistent with the evidence strewn about these once-dominate American companies is that they all had massive, top-down control systems. Therefore, those control systems are at the root of American business failure.

We are amazed that those appeals had—and continue to have—such a wide following in the face of continued and colossal failures of companies led by managers following that prescription for profit. Few people questioned whether the apparent failure of the discredited control systems was caused by poor application of a sound theory. A similarly constructed argument would condemn Christianity or Islam because people claiming to act according to those religious precepts behave atrociously. Neither argument is logical.

The arguments supporting pro-chaos fads disintegrate when we admit that business systems are artificial systems. They are the antitheses of nature. By intention and design, they remove as many of the forces of nature as possible to improve our economic well-being. The history of Western business has been and will continue to be a struggle to form order out of chaos and use that order to build wealth. What makes pro-chaos faddism the grave of American business is that so many people have mistaken foolishness for wisdom.

If overcontrol is not the common source of business failure, what is? The root cause of all business failure in the Middle Ages as well as into the 21st century is uncontrolled change. All commercial failures can be traced directly to inappropriate uncontrolled change in an entity's systems. It goes like this every time: Management has ample internal and external sources of information disconfirming the current methods of doing business. They refuse to accept that information and therefore fail to act on it, because they lack the needed fundamental skills. Lacking those skills, they lack the security to move forward. Remember, to accept disconfirming information, people need psychological safety, the feeling that they can handle bad news because they are confident they can find a way out. Instead, management hires someone hawking a canned management fad, which only camouflages the problem.

That is not to say that catastrophic economic failures do not occur in spite of well-constructed, dynamic operating systems. Catastrophes can shut businesses down even if management does a superb job of contingency planning. For instance, a fire in our data system may give our competitors a window of opportunity. An earthquake in California could immobilize our California brokers. Our investment in Labequam could be nationalized. Planning for such possibilities is prudent management, but no matter how much planning you do, you still could be overwhelmed. Just because you know that chaos may dictate the outcome of a future event is no reason to give your life over to the Dark Force. After all, in the end, we are all dead. Nevertheless, we overcome that ultimate and very personal chaotic event by making plans to improve ourselves in the brief time we are here. Likewise, it makes little sense to turn our productive business lives over to chance just because the cosmos operates on a random-number generator.

Finally, our theory that all business failures can be traced to inappropriate uncontrolled change has to explain activity at the margins—corporate death by natural causes, also known as unforced liquidation. Liquidation is a failure only if it is unplanned, and unplanned liquidation comes about because of inappropriate uncontrolled change of a company's systems. Conversely, planned liquidation or restructuring can be the most logical and fruitful action management can take, given that it is in response to an accumulation of facts. Accepting disconfirming information pointing toward liquidation is a difficult step for most managers. Nonetheless, even the best run company can find itself on the wrong end of a bad bet that must be covered. A management team that understands the fundamentals of controlled change fully debates and understands such bet-the-company risks before undertaking those risks. Since liquidation can be a planned response to a risky outcome, such liquidation is not, per se, the mark of failure. Our claim that all business failures result from inappropriate uncontrolled change still holds.

6.5 THE RELUCTANCE TO WRITE POLICY

If controlled change is real, it has to occupy physical or mental space and leave a trail. The primary evidence that controlled change exists is the existence of company policy. Policy, whether written or just understood, cradles the proven and essential rules for operating the company successfully over a long period of time. Those rules embody the institutional memory and have stood the test of time. Many times, they were established or expanded when the company suffered episodic near-catastrophes. Suppose you want to change the way something is done in your company and the anticipated change violates policy. By definition, you are talking about something the company's upper management and institutional memory consider substantial and significant. Otherwise, your change would affect only those procedures that upper management leaves to you anyway. When your idea is implemented, the company's policy will have to change to accommodate your idea. That change process represents the role of policy making: to ensure that significant changes are brought to the attention of appropriate people. How policy is changed separates good systems from bad.

Whenever we begin an assignment with a company looking to increase profitability, the first tool we help develop is an up-to-date operations policy manual. Typically, upper management pawns the responsibility for writing policy on some unfortunate middle-management person. With such an inauspicious start, the project seldom gets anywhere until the middle manager finds a way to reengage the boss. Why is it so hard to get upper management to do its job in setting policy? And why is policy setting basic to improving profit?

By now, you know the answer to why upper management resists writing policy. Writing a policy manual or updating an old, forgotten one represents significant change. Any change is difficult, especially one that exposes people to great uncertainty and provides little safety. Besides being leery of the consequences of systematic attention to policy writing, top managers do not want to write policy because they are uncertain where to start, how to write it, why it is needed, and when to stop. Should they finish writing a manual, they have no idea what to do with it anymore than their subordinates do. But the fact remains that every company has a multitude of policies on everything from how long to hold accounts payable to how many vacation days can be taken to who gets a key to the office and a parking space. Unfortunately, the only time policy is written down is after a problem arises from its being unwritten.

Another reason general managers are reluctant to write policy is that they feel that if they reduce policy to writing they will lose flexibility. All managers believe that their company's business is different, and to a great extent they are right. When asked what makes their business unique, they usually point to intangibles: the maturity of their work force, the quality of their product, the superiority of their management. Up to this point, they may have operated with policy manuals that cover personnel issues, but they have had no reason to write a manual covering operational aspects.

That is the very reason they need to write one. They are ready to take the next step toward improved profitability, and a policy manual is the first step. A policy manual builds the psychological safety necessary to get the management team to move forward.

Last, some top managers resist writing operating policy because they have turned their company over to their lawyers and do not even know it. In the case of writing policy, some attorneys believe they can see into the future. They will tell you that you should avoid writing policy because it could come back to haunt you in an as-yet unfiled but conjured lawsuit. In a later chapter on decision theory, we will show why that kind of "scenario thinking" is bad technique. For now, let us just accept that any advice given by any specialist concerning business practice has to be examined thoroughly by management experts, that is, by people like you. In the case of advice from attorneys concerning policy, use it to help you reduce risk; however, never let an attorney tell you that you should not have a policy manual. If you find yourself getting business advice too often from an attorney, get a new one who spends time helping you get something done instead of giving you reasons why not to do it.

6.6 PLAN, POLICY, AND PROCEDURE

While we might understand management's reluctance to write policy, we will not be able to overcome that resistance unless we build a strong argument of how a well-written, cogent policy makes money. We need to explain why policy is crucial to profit improvement and why written policy is crucial to sustained improvement. To do that, we have to build a model of how all businesses operate. While we do that, keep in mind that this model is essential to understanding how change occurs in companies and therefore how policy functions as a change tool.

But before turning to business systems, let us look at an analogous system, a typical family. Families have policies and procedures. Typical policy statements might be "No child goes on an unescorted date before age 15"; "Every child will go to college or trade school"; "We marry within our faith and race." Few of these policies are written down, but every family member knows that to change a policy requires a persuasive argument.

Suppose that one of the children takes a fancy to someone of a different faith or color and is entertaining marriage or that another child decides to quit high school and join a band. The consequences of those decisions go to the root of what the family had accepted as appropriate behavior. Those types of decisions require a complete review of the policy by the "managers" of the family. Father, mother, and maybe the grown children will debate them, and those wanting change will argue vigorously. Maybe the policy will change, maybe it will not. But there is little doubt as to the importance attached to the decision. It will affect what it means to be *this* family. As for the procedures that implement family policy, they might include setting curfews for the children, requiring meeting the children's new friends and their parents, and setting aside certain times of the day for study. If one of the children habitually disregards a procedure, the reaction of the parents and the family as a whole is less intense than if the child jeopardized basic policy and values. An example of a written document backing up family procedure is the chart on the refrigerator door that lists everyone's whereabouts and weekly chore assignments.

The analogy of policy and procedure in a family fits well with business systems. While businesses are not families, they are collections of people who need to know the basic rules of conduct and operation. The first ingredient in any business operating system is the hierarchy of plan, policy, and procedure. Later we will make a distinction between procedures and instructions, but for now we will lump them both into procedures. All businesses have them, and you will realize that the distinctions are valuable tools to controlling change. Also for the moment, let us put aside the role of documentation in operating a company and assume we are looking at a pretty typical business that is marginally profitable.

At the highest point of any business operating system is its plan. While the business plan might be written, many times it is not. Whether written or just dwelling in the company's collective subconscious, the plan contains your assumptions about your market and your place in it. You have certain basic assumptions about the content of your base technology, your pricing relative to your competitors, the valuable features of your product or service, and your competitive advantage. Your plan sets the boundaries of your business. For example, suppose your primary product is made of plastic. You might define yourself as a specialty plastic molder or a commodity plastic molder. Or you might make no distinction about the process but concentrate on the product instead. You may believe that your competitive advantage is in cost-effective production, unique product features, or incredibly fast turnaround. Such elemental decisions are the type that reside in your plan.

Your *operating policy,* on the other hand, contains statements about what you are going to do to implement your plan. These "what" statements purposely avoid mentioning who is responsible for executing the policy or how it will be done. Suppose your plan emphasizes your competitive cost advantage. Your policy then would reflect that focus. For example, your company might have the following operating policy statements:

- Each and every incoming order is reviewed for opportunities to reduce costs. The results of the review are documented.

- Pricing is the responsibility of one department; its decisions are final, documented, and reviewed by management on a quarterly basis.
- Product development will focus exclusively on cost reduction. New-product innovation is the province of our customers, other higher-priced build-to-order competitors, and the marketplace.
- Distributors are relied on 100% for selling our product. Distributor contracts are reviewed and approved according to written procedures. The responsibility for distributor relations is assigned to one person (and staff, as appropriate). Distributor performance is reviewed and documented twice a year.

Whether policy is written or not, it still exists. Even if the policy shown in the preceding list were unwritten, you still would find evidence of it in the documents the company generated. The order files would have a sign-off from Industrial Engineering about their review of cost reduction, and there might be a fax to the customer asking for a variance in the specified requirements in order to reduce the price 30%. Over in Accounting, you would find records showing price and cost reviews. In the president's office, you might find all the distributor files and in them checklists documenting the review and approval of the contracts. You also would find records of the semiannual distributor meetings held at a downtown hotel. The fact is that the company has an operating policy that reinforces the business plan.

Procedures, on the other hand, tell you *how* the policy will be put into place, not what it is. They implement the policy. Continuing with our example, let us look at how the distributor relations policy might be implemented. The fourth bulleted item in our policy list says that the contracts are to be "reviewed and approved according to written procedures." The checklist we found in the president's distributor files is the written procedure that the company follows when it reviews a distributor contract. That is an important distinction between policy and procedure and has a profound effect on how well behaved your change control system will be. The policy says *what* was going to be accomplished, that is, that the review would be according to a written procedure, but it purposely does not say *how* that review is to be done. Look at what the policy says about reviewing distributor performance. It says that it will be reviewed and documented twice a year. It leaves the content of the review and the form of the documentation to the procedure. Again, this is not a trivial point, as you will see later.

6.7 THE EFFECT OF DISCONFIRMING INFORMATION ON OPERATING SYSTEMS

Earlier we made reference to internal and external disconfirming information without defining our terms. Now those distinctions will become clear.

Figure 6.3 illustrates the effect of information on the company's plan, policy, and procedures. It distinguishes among three sources of information and how much may be necessary before your procedures, policy, or plan changes. The horizontal axis plots the relative amount of information being absorbed by the company. The vertical axis illustrates different thresholds for change. Anytime a line crosses any one of the three thresholds shown (plan, policy, procedure), that part of the control system will change.

In Figure 6.3, the line labeled A shows how internal information might influence a company's control system. It takes a considerable amount of internal information before procedures begin to change, but at some point enough information is readily available that the annoyance of working around a problem brings people together, and a procedure is changed. For our purposes, it is immaterial whether this is a planned change or not. (In fact, we have yet to define what a planned change is.) As a way of reinforcement, let us again revert to our sample policy and procedure list. Suppose the sales department continual finds skimpy historical information on the incoming-order cost review. This internally generated disconfirming information indicates that the cost review procedure is not working well, at least according to sales. Up to this point, the industrial engineers (IEs) have been solely responsible for the review. The sales department's solution to the problem of insufficient documentation is for them to be included in the cost review process. They do what they have to do to be included in the loop. The

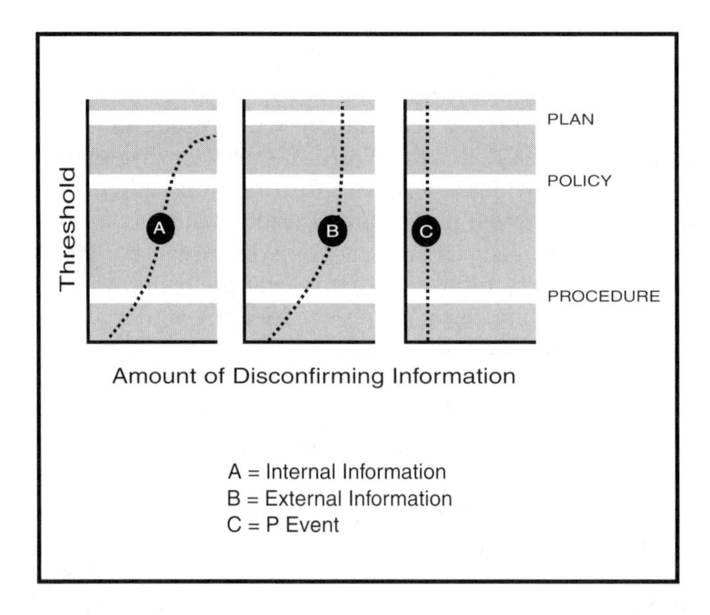

Figure 6.3 Change threshold model.

unwritten procedure is changed, and maybe a new form is developed, maybe not. Life goes on, but the procedure for incoming-order review has changed.

Now suppose that cost data continue to be inadequate as far as the sales department is concerned, while the IEs are concerned that sales' new role in cost review is making the process far too slow and costly. At some critical mass of internally generated disconfirming information, a policy change has to be made. This occurs in Figure 6.3 where line A crosses the policy threshold line. Note that the stated policy says nothing about *who* is responsible for the cost review; it just says that each order is subject to one. As a rule, you do not want to include responsibility statements in policy. To be effective, policy statements must be fairly stable, while work assignments can and do change, more often than you think. If you make too many responsibility statements in your written policy, you will constantly be issuing new policy revisions, that is, if you have a written operating policy.

Now you are gaining some respect for the reasons why we make distinctions between procedures and policy and why writing both of them has such value. Without a written procedure, no one will know whether the role of IEs as sole cost reviewers is a convenience of procedure or an artifact of policy. Does general management think that IEs should be the only ones in the loop? If so, that is a policy statement because it is a constraint with the force of policy. Policy is axiomatic thinking that is not easily changed; when it is changed, general management has to give its consent. If management assigns responsibility at the policy level, they are setting up a semipermanent gatekeeper. It usually means that the company has learned, over the years and the hard way, that a particular part of operations demands an unequivocal statement of responsibility. On the other hand, if general management believes that cost review should be a variable function that depends on the best currently available staff and ideas of the day, then they will keep responsibility assignment out of policy and instead leave it in the procedure domain.

Sometimes information overwhelms policy, and the business plan has to change. That event rarely occurs when internal information is the only driver. Insiders consider internal information as selective, full of built-in bias, and weighted with the hidden agendas of different departments. Even general management is reluctant to change business plans based solely on repetitive and overwhelming internal information. That is why Figure 6.3 shows internal information effects (line A) topping out before it reaches the plan threshold.

External information works the same way as internal information, but its effect on operational systems is quicker and more dramatic. External information by definition comes from outside the company even though an insider may present it. For example, suppose the sales department reinforces its internal information for better cost review with the fact that higher-priced competitors are now beating the company on price alone. More functions will be interested in this new external information; no longer is it an intramural issue. The marketplace is offering up disconfirming information. If similar information continues to come in, procedures in many areas will be reviewed to ensure that the company is doing all it can to remain the low-price leader.

At some point, procedural changes are just window dressing if what you really need is a policy change. Suppose the root cause of losing orders on price is not your cost structure but distributor relations. Suppose external information says that your big distributors are getting into the production business, and to do so they have cut prices of their own lines to the bone. At this point, you need to review your policy of relying 100% on distributors. You may decide you need a policy concerning distributor-manufacturers. Or you may decide that you need a mixed distribution system in geographical areas where distributors are quasi competitors. Whatever your response, changing procedures will not solve the problem. Your policy has to change.

Another inherent function of written policy is now evident. When you are collating information and looking for the root cause of a problem, you need a way to categorize the seriousness of the problem. In the preceding example, pricing pressure in the marketplace is an obvious problem for top management. Or is it? It might be obvious to top management that they should be included in the loop, but what if it is not obvious to the people who actually get the information? Remember how Vern empowered himself to add a product feature (lift eyes)? Because Vern's company was unclear about how information should be classified and how changes should be made, Vern was not able to tell the "right" people about his problem because the right people were not identified. The right people may not even have known that they were the right people. The power of making a distinction between policy and procedure is that if the root cause of a problem requires a policy change, you know you are working with a significant issue.

At some point in the change threshold model, line B (external information), unlike line A (internal information), crosses the plan threshold. Over time, we may get irrefutable information that our business plan is suspect. We have changed procedures and we have rearranged our distribution channels, but we still are losing price leadership. Management has got to ask if the basic assumptions about the business are valid. If we change the business plan, then we have to revisit our policies and procedures. That is why categorizing an operating system along the lines of plan, policy, and procedure is so robust. The model demands discipline and consistency between levels. While not everyone may not be cognizant of those distinctions, they necessarily are affected by them.

The last classification shown in Figure 6.3 is line C, called the "P-event" (for all the Ps in the model: plan, policy, and procedure.) The source of

P-information is almost always external, although accompanying internal disconfirming information may have been building for quite some time. Nonetheless, dramatic and undeniable information tells management that the world has changed and that every part of the company's operating system has to change. In the case of a P-event it does not take a preponderance of information to force the issue, and the first place management effort needs to be applied is at the plan level. Change in policy and procedures will follow. Some recent examples include oil prices at $40 a barrel in 1982 and $10 a barrel in 1987, passage of the North American Free Trade Agreement, collapse of the Mexican peso, reunification of Germany, and liberalization of the Chinese economy.

6.8 PLANNED AND UNPLANNED CHANGE

So far, we have made little distinction between planned and unplanned change. All companies have both explicit artifacts and implicit understanding of this operating model. All companies change in about the same way because the change dynamics introduced in Chapter 4 exist in all companies. But those companies that plan for change have a plan! That is, they must have a systematic way of changing their plans, policies, and procedures. To put emphasis where it needs to be, they have procedures for changing procedures. Everyone in those companies knows who is responsible for what work, and everyone knows when line A, B, or C is changing. Those routines ensure that appropriate levels of management are aware of procedural changes without excessive oversight while allowing appropriate personnel to be intimately involved when the situation demands it. Those companies accomplish that by sufficiently and explicitly defining their operating systems such that change opportunities systematically find their way to the right people.

We started this chapter by stating that all business failures could be linked to inappropriate uncontrolled change. It is not that uncontrolled change has to be avoided at all costs. A certain amount has to take place or you are in danger of overcontrolling, that is, spending inappropriate attention and money on identifying, documenting, and reviewing disconfirming information instead of relying on the efficiency of natural change. So how do you identify inappropriate uncontrolled change? When procedure, policy, or the business plan is changed without management reviewing the underlying disconfirming information. Whenever you cross one of the thresholds in Figure 6.3 and no systematic review of the changed plan, policy, or procedure involved takes place, then you have inappropriate uncontrolled change. Management's responsibility is defining those three thresholds and building a system that reinforces them. That is the essence of controlling change in business. Now you know why effective use of policy and procedure is one of the fundamental manageering skills introduced in Chapter 1. Without that fundamental skill, you will not be able to use its companion skill of controlling change. Chapter 7 explains tactically how you can build a system of operating policy and procedure without having to reinvent the wheel.

6.9 EPILOGUE

While we have built the case for written policies and procedures, some extremely profitable companies have very little documentation. But those are rare birds. Instead of explicit policy and procedures, they have exceptionally consistent and incredibly driven upper management, usually in the person of an owner or a significant shareholder. Such people are, by definition, freaks. Their powers of analysis are superb, and their understanding of policy and procedure is second nature. They intuitively know the difference and marshal their assets accordingly. They are the teachers to the rest of us. But most of us are nowhere near their equals, no more than we can defend against a Hakeem Olajuwon fake or hit a Nolan Ryan fastball. We have to use tools to make up for our shortcomings. To ignore that fact is to reduce the potential for profit and to lose your opportunity to have enduring impact on the companies, divisions, departments, and people who look to you for leadership.

Chapter 7

Documentation Basics

Chapter 6 argued that managers must appreciate the central role that policy and procedure play in running a business if they are going to have any chance of directing change and thereby sustain profitability. We also made the case that every company's operating system is anchored by the company's policies and procedures, and what distinguishes one company from another is how well its policies and procedures handle systematic change. But while the change-threshold model illustrated the drivers that compel a company to change its plans, policy, and procedures, the model said nothing about how someone might go about building and cultivating a healthy system that takes advantage of the model's insights. That is what this chapter does.

After you understand the nature of change, the remaining key to *controlled change* is understanding the nature of control. The only way you can control an organization's operating system over the long haul is with documentation. The resulting documentation scheme must produce valid, pertinent directions and action plans when and where required without undue cost. But instead of presenting you with a model of an efficient document-maintenance system, we first need to help you appreciate the much maligned concept of hierarchy. Then we will discuss some basic elements of documentation, including the introduction of document-system efficiency. Only then, when you have a basic understanding of the "why" behind business-system documentation, will we be able to successfully promote the "how" of a particular documentation strategy. So for the moment, relax, sit back, and gather a valuable historical perspective.

7.1 THE PARADOX OF CONTROL AND EMPOWERMENT

The relationships described in Chapter 6—between policy and procedure and the role management plays in establishing and maintaining them—is a familiar business model that has changed little over the past 100 years. Take the standards-based management-systems fads of the 1990s. Behind them was the assumption that the hierarchical model of responsibility and authority is common to all companies. Those written operating standards, created and supported by various industrial and professional organizations, used the hierarchical model as a tool to compel target companies to document certain basic operations so that they could be audited against the requirements of the standard being considered. Like all business fads, that one promotes one element of the enterprise over all others, but it had a different twist.

Most fads of the 1980s and 1990s stressed, in some form or other, the value and righteousness of flat, presumably nonhierarchical business units. The standards-based management-systems movement, born among the waves of this enlightenment, did not fit that trend. The presumption that every business is organized along hierarchical lines (the so-called command-and-control structure) reflected its heritage. Those standards were cousins of earlier military procurement standards. That dubious family tree made some people uneasy and others downright hostile toward standards-based management systems since the military model of management was dismissed as inappropriate for the private sector. (As of this writing, standards-based management systems continue to grow in spite of that concern. We discuss the problems and value of those systems in Chapter 11.)

The rush to protect the newly empowered working man and woman by burying command and control in favor of laissez faire management ignored two facts. First, command and control and employee empowerment are not mutually exclusive. Second, regardless of how flat an organization might be, hierarchial relationships are necessary for work to get done. Just because we refuse to draw a hierarchical organization chart does not mean that the "real" organization has no such tendencies.

For instance, in the early 1990s, Intel, a leading manufacturer of microprocessors, was featured prominently in the popular business press. Intel was a wildly successful company, and much was made of the role of the company's "culture" in that success. The press was fond of pointing out that Intel's offices were open, allowing free access to anyone at anytime, thereby fostering all sorts of efficient communications and egalitarian feelings between management and the worker bees. We were breathlessly told that even the CEO had his desk in such an open-space environment. This office arrangement supposedly was the paradigm for good business organization. The story matched nicely with the popular management fads of the day. When Intel suffered an embarrassing production problem with a popular processing chip in the mid 1990s, the bloom came off the rose. The same business press revealed that the CEO seldom if ever used his open-space desk (gasp!). He was, according to updated press reports, more a taskmaster closed to other people's ideas than a benevolent leader open to contrarian views.

We tell this story not as an illustration about Intel, but as an example of how easy it is to mistake form for substance when discussing how work gets done. Whether Intel's CEO was what the press says he was or whether Intel's success and subsequent problems were related in some way to its "culture" and management style is beside the point. The point is that this type of patterned thinking (nonhierarchical = good; hierarchy = bad) obscures a basic truth of group behavior and business organizations. Regardless of whether your management style is authoritarian or egalitarian, the people working in your company require *some* level of command and control to operate efficiently. Command and control does not mean that every aspect of the company's operating systems are described in detail. It means that the boundaries between responsibilities and authority are well established and communicated. For example, it is obvious that the accounting department is not expected to write a law brief on the latest environmental statute, nor is the engineering department going to cut payroll checks. Where those boundaries are set and, more important, how they are changed is the province of command and control and is absolutely necessary in a well-run company of whatever management flavor.

7.2 WRITE IT DOWN

While command issues can be intellectually separated from operations issues, for our purposes we make no distinction between the two. Our working assumption is that most companies do not distinguish between the two since they rarely discuss either. We have established that the question is not whether you have methods for controlling operations, but how well those methods work, especially when under attack. But what distinguishes a good system from a bad one?

First and foremost, to be of any use, an operating system must be written. Documents are the primary means for understanding, illustrating, and communicating how work is supposed be done. While it may be obvious that writing is a good communication tool, we tend to forget that many times the resulting document is not as valuable as the knowledge gained while composing it. Francis Bacon observed that "writing maketh an exact man," meaning that writers are forced to develop a solid grasp of their subject while speakers are allowed to be less precise. Besides forcing us to be more disciplined when explaining ourselves, written operations documents serve as a record of our "best methods" at the time of writing. Many operating documents are useful only when we are trying to understand why something went wrong. Without the benefit of a record of how something was supposed to be done, we have no idea whether our current practice matches our best practice. Furthermore, if we believe that something needs to be changed so that we do not make the same mistake, without documentation we really do not know where to start. Everyone will have a different recollection of what the "current state" was or should have been.

7.3 TOO MUCH IS AS BAD AS TOO LITTLE

Just because a system is written does not necessarily mean that it is a good one. Documentation is necessary but not sufficient. Such a system could be bad if it either (1) generates excessive, redundant paperwork or (2) fills up everyone's credenza with thick manuals that just gather dust. These predicaments are not inherent to document systems themselves but instead reflect their poor construction and implementation by people who have little training building and maintaining the systems. The assumption that if you can write you can build a good document system is equivalent to believing that if you can speak you can play Othello. The only way you get to play Othello is by learning the fundamentals first. Your first task for documenting your management systems (or improving the one you have) is addressing how the documents themselves will be established and maintained, not what will be in them. The document system we suggest later will suggest ways to do that. The resulting system will thereby control the tendency for document systems to crash of their own weight.

7.4 DISINTERESTED MANAGEMENT DOOMS THE SHIP

Whenever the basic operating philosophy and core documentation procedures are created or changed, top management must be involved in more than just assigning the work to some function and signing off on the work. Building or revitalizing a company's document system, whether paper- or electronic-based, is like operating on a vital organ. General management must take an active role to understand the ramifications of such major surgery on their company. Otherwise, they have no right to complain that the patient is not getting better.

7.5 A DOCUMENT BY ANY OTHER NAME

There are apparent differences between paper-based and electronic-based document systems. For the most part, however, how you design a document system is independent of the physical form the actual documents take. For example, a document *system* exists whether or not actual physical documents exist. The concept of a document is so indispensable that electronic-based information systems continue to refer to their internal records and output as "documents" even if they do not produce a physical piece of paper. A document is only a tool by which we store information or evidence. It is a record of something in the past. The immutability of ink on paper is a required feature of document systems, and computer-based systems try to emulate that by controlling who can make changes to its document templates and how those changes are archived.

7.6 SELF-REFERENTIALISM

A document system is a peculiar animal, which is why it is so easily abused. Much of a document system's documentation is devoted to describing the system itself, a trait called self-referentialism. We encounter self-referentialism everyday in other parts of our lives without giving it much thought. It is second nature. Whenever you talk to someone else, you describe bits of the world by using your own personal reference. You unconsciously assume others have had comparable personal experiences. For example, when you refer to "quality," you refer to an ideal drawn exclusively from your own experience. Self-referentialism surrounds us—quality, beauty, love, pain, happiness—and can cause serious misunderstanding, but we seem to muddle through our daily lives in spite of this permanent condition. However, muddling through self-referentialism in document systems eventually cripples it.

Self-referentialism explains why it is so easy to chase your tail when you do some quality (there is that word again) thinking about document systems. It also explains why people who work on and around those systems everyday seem a little goofy. You would be, too, if you struggled with paradox every day. Self-referentialism also helps us understand why document systems have an inherent tendency to become unstable, that is, they themselves start generating paperwork. It also helps us understand why having procedures for writing procedures is essential to maintaining control over procedures. (If that statement is not self-referential, nothing is.)

7.7 DOCUMENT SYSTEM EFFICIENCY

To understand how to build (or rebuild) a document system, we need a way to describe something we call its efficiency, that is, a measure of how much the system itself is disrupted when a covered event in the real world has to be recorded. While the following discussion can be applied to any document system (e.g., financial, environmental, legal), our examples will be directed at a company's operating system documentation.

The following attributes determine the efficiency of a document system:

- *The number of people who get a revised document when job content changes.* For example, suppose the drafting clerk adds a new step to the weekly file maintenance checklist. Is the clerk the only one who gets a copy of the revised checklist? Is it copied to the whole drafting department? Do other departments get a copy, too?
- *The number of documents that are affected when job content changes.* In the preceding example, is the checklist the only document that needs to be changed? Are there master indexes that need to be changed? Are there other procedures that need to reflect the new step?
- *The number of people responsible for maintaining any particular document.* Continuing the example, is the drafting clerk the only person responsible for maintaining the checklist? Does the proposed revision need to go through other channels? If so, how many people are involved in looking over the change?

If your document system rates low in all three attributes (few people receive revisions, few documents are affect by change, and few people are responsible for any particular document), then you have a well-behaved document system. Surprisingly, the absolute number of documents in the system has nothing to do with the document system's efficiency.

These efficiency attributes apply to paperless document systems as well. In fact, these efficiency measurements may be even more important in paperless systems. It is easy to suppose that computer-based document systems are easier to maintain since their records are readily changed and have virtually unlimited capacity. But as most people who work with large databases will tell you, what keeps them up at night are breakdowns in the disembodied information model that describes the relationships between the data in the machine and the real world, not changes that obliterate information. They can usually resurrect obliterated information from the redundant archives they keep. They cannot rebuild obliterated document relationships. Computer-based relational models cannot keep up with unstructured, mindless change anymore than paper-based relational models because the model resides outside the database itself. Just because computer-based document systems can keep up with document changes more efficiently than paper-based systems does not mean that they can keep up with turmoil in the real world better than paper-based systems.

That points to a central difficulty that designers of computer-based document systems have that designers of paper-based systems do not. They have to make managers and users understand that while the system can make information available to anyone and everyone, it is not necessarily in the best interest of a well-behaved, sustainable system to do so. Computer-based document systems may make it easy to increase the number of people who are informed of a change in the drafting clerk's routine, and those systems may make it easier to pass the proposed change by more people for review than a paper-based system. But doing so actually reduces the document system's efficiency because it increases the amount of low-value work going on in the company. This is a classic case of more information actually reducing overall value. In practice, many of the people who are added to such an electronic distribution list could care less about the information, but the system requires that they spend precious time discovering the fact they could care less. That is why the document efficiency attributes we presented are independent of the tool used to maintain the actual documents. What is vitally important is that people's time is used efficiently, not that excess information-system capability be put to use.

7.8 "FOR YOUR INFORMATION" USUALLY MEANS OTHERWISE

The first efficiency attribute is the number of people who must get copies of revised documents (or be notified that a change has taken place). While it seems obvious that you need to control the number of people on the distribution list, how to do that may not be so obvious. First, you have to refrain from distributing paper for the purposes of "keeping people informed." Just because we can copy or e-mail the world does not mean we should. We have seen simple clerical work instructions distributed to more than a dozen people. Most of the people on the distribution list were not really interested in keeping such close surveillance on the work processes of a single person. On the other hand, should two or three of those dozen want the information, a well-behaved document system should help them find the information when they feel the urge. One way to combat some people's urge to send copies of trivial procedural documents to the entire company is to have a document policy that restricts distribution of any document to people actually mentioned in the document itself.

Even if you restrict distribution to the people mentioned in the document in hand, your system still may be a paper generator because of an obscure, structural problem: documents with too wide a scope. The more people included in the body of a document, the wider the required distribution. The more activities discussed, the greater the likelihood of duplicate and conflicting information. For example, suppose the clerks in order entry are now required to date stamp each order when it hits their desk. If that change requires distributing the revised instruction to the entire department or, worse, to people outside the department, you have written a document that has too wide a scope. That usually happens when instructions are imbedded in documents that ostensibly describe how procedures work. Furthermore, if you have a tendency to describe too many activities in one document, the chances that another document also describes the same activity increase. When an operational change occurs that affects repeatedly described work, revising the redundant documents is more difficult than it would be if you severely restrict the scope of each document.

To put the brakes on the natural tendency to expand rather than contract document scope, you need a policy that limits the length of procedures and instructions to five double-spaced pages with 1-inch margins and a 10-point font or an equivalent electronic limitation. You have to be *that* specific because people will start playing with margins, spacing, and fonts to keep within a simple page limit. Should a document begin to grow past the five-page limit (and they always do), it should be broken into two more focused documents.

Counterintuitively, getting a good score on the "number-of-people-getting-a-revision" attribute usually calls for more, not fewer, documents. That means your document control system is essential to ensuring that the system does not spin out of control. We will discuss how you can do that after we address the remaining efficiency attributes.

7.9 DO NOT NAME THE NAMELESS

How can you limit the number of revisions required when someone changes jobs or when job content or responsibilities are rearranged? A documented operating system that needs major revisions when confronted with ordinary business changes is suffering either from being too specific in describing hand-offs or from a tendency to write documents around people. (A hand-off occurs when data or physical items are handed off from one department to another across a predefined process boundary.) In the instance of being too specific with hand-offs, suppose your order entry procedure uses job titles in describing how order entry works. That is excellent practice to portray activities within the function under scrutiny, but if that same document continues to use specific job titles in describing hand-off activities, a future filled with necessary but avoidable revisions is guaranteed. Also, the revision process becomes much more difficult. Changes in work on one side of the boundary now have to flow back across and show up in multiple documents on the other side, causing an unbelievably chaotic and tangled mess.

To avoid that systemic weakness, restrict job titles to activities within the function being described and use department or functional names when describing boundary conditions and hand-offs. This documentation discipline has to be outlined in your procedures for writing procedures and reinforced by vigilant management. It is dull and sometimes tedious, but nothing works as well as upper management returning a procedure to its owner with a note concerning poor documentation practice.

Being too specific when describing hand-offs is not the only way a document system becomes difficult to revise. Even if you are careful to avoid specificity when necessary, you still may generate too many revision requirements if you write documents around people instead of processes. In most companies, some individuals hold several distinct jobs concurrently, evidenced by the number of people who say they "wear many hats" or that their titles do not describe what they actually do. At the high end are the CEOs who also chair their boards. At the factory level, a production scheduler might also be the quality control manager but hold any number of titles—scheduler, QC manager, expediter, vice-president. If those autonomous jobs are lumped together in the documentation into one meaningless job title, the document system will turn inside out when the company expands and contracts. Continuous expansion and contraction, analogous to breathing, make you split, combine, or rearrange jobs constantly. Your document system must be stable during ordinary life processes if you want to be able to take advantage of your environment. You cannot hunt for elephants if you are on the operating table all the time.

You can avoid turmoil in a document system by building the system with an eye on job descriptions, not on the actual people who hold those jobs at the time. If a person holds a job that really encompasses several distinct functions, split the job into as many stand-alone positions as necessary. You can discover those multiple personalities by using the free-form organization charts introduced in Chapter 5. When the company expands and people are hired to take over some of those responsibilities, the document system will hardly notice. When the company contracts, all the positions will still exist, but several may be filled by the same person.

Another helpful technique when you are designing or repairing a document system is to ask yourself, "What happens to this procedure if the number of transactions doubles or is cut in half? How do we structure work and its controlling procedures to ensure that we do not have to redesign everything each time the company expands or contracts?" Should the company's sales double or be halved, your goal is for the document system to be essentially unfazed. At some point, however, significant document revisions will be necessary. When the anticipated changes are no longer at the procedural level but instead pierce the policy threshold, it is time to overhaul the system.

7.10 TOO MANY COOKS SPOIL THE BROTH

The last efficiency attribute examines the average number of people responsible for maintaining any one document. A poor showing on this attribute—more than one person, on average, owning any one document—sometimes reflects confusion between *system* maintenance and *document* maintenance. System maintenance is an administrative function that ensures that documents are distributed in a timely fashion to the appropriate personnel; that the latest revisions are available while obsolete ones are withdrawn; and that everyone is aware of and follows the rules, like adhering to procedures for generating procedures. Document maintenance, on the other hand, is concerned with ensuring that the actual information in the documents themselves is the best, most up-to-date possible. Those activities—system administration versus document maintenance—serve entirely different purposes. Like the people who wear more than one hat in a company, those jobs may rest in the same person, but they are two very different, distinct, and independent responsibilities.

The easiest way to hobble a system is to make no distinction between system administration and document content. It might take time for the document system to crash, but eventually that flaw will be fatal. On the other hand, if you are looking for the quickest way to hobble a system, be sure to habitually assign individual document maintenance to more than one person (i.e., give two or more people the "joint responsibility" for the content of a document). That usually happens when you are reluctant to decide who should be responsible for the outcome of a joint activity. Common but deadly joint responsibilities are making sales and accounting responsible for order entry, production and engineering responsible for design control, and sales and production responsible for scheduling.

The way to avoid such fatal problems is to expect one function, like accounting, production, or inspection, to be responsible for overall administration of the document system but to assign ownership of individual documents to individual job positions. It works like this: The system administrator keeps the master list and master copies of the documents, trains everyone on how the document system works, and periodically audits how well it is operating. Each document, when first written and distributed, is assigned an "owner," who is responsible for keeping the intellectual content of the new document current. If someone other than the owner has a problem with a document (e.g., is not able to find needed information or believes that a revision is required because of a change in work flow), he or she goes to the owner of the document, not to the system administrator. The owner is responsible for reviewing the proposed change, checking with related functions at the border of the affected procedures, making the required changes, getting appropriate approvals, and submitting the revision to the system administrator for insertion into the system. By separating system maintenance from content maintenance, system administration gets the attention it needs without being caught up in technical content. By restricting document ownership to the single person with the most vested interest in the procedure, you make sure content gets the attention it needs to remain up to date.

At this point, you may be absolutely beside yourself, shouting back at this book, "What do you mean I can't have joint owners of procedures? It is inevitable that the most important value-added procedures have to be owned by several people! We are a team here!" If you skip a couple of pages and read Section 7.13 we answer that objection. Just remember to come back to learn more about designing a document system.

7.11 THE DOCUMENT PYRAMID

The rules for a document system help you get a good system efficiency score but omit any reference to a method for keeping track of the documents themselves. That is where a document classification framework comes in. In Chapter 6, we discussed the difference between policy and procedure and explained how the distinction is useful for understanding how change occurs. We mentioned that some procedures are actually work instructions but did not elaborate on the esoteric difference. We now return to that additional differentiation.

The document pyramid shown in Figure 7.1 is the universal tool used to organize well-behaved, unit-based document systems. Along with other ideas presented in this chapter, the document pyramid forms the basis of most successful document systems regardless of whether we are talking about financial systems, operating systems, air-traffic control systems, welfare systems, or the legal system. This framework categorizes a document by type, ensuring that its scope remains focused and consistent with its place in the system that maintains it.

The difference between procedures and instructions is as important as the difference between policy and procedure. *Procedures* tell the reader how policy (as described in the policy manual) is implemented. *Work instructions,* on the other hand, define how one particular job position (person) works within the confines of a procedure. These are *two* different documents. Suppose you have a policy to ensure that you understand what a customer wants

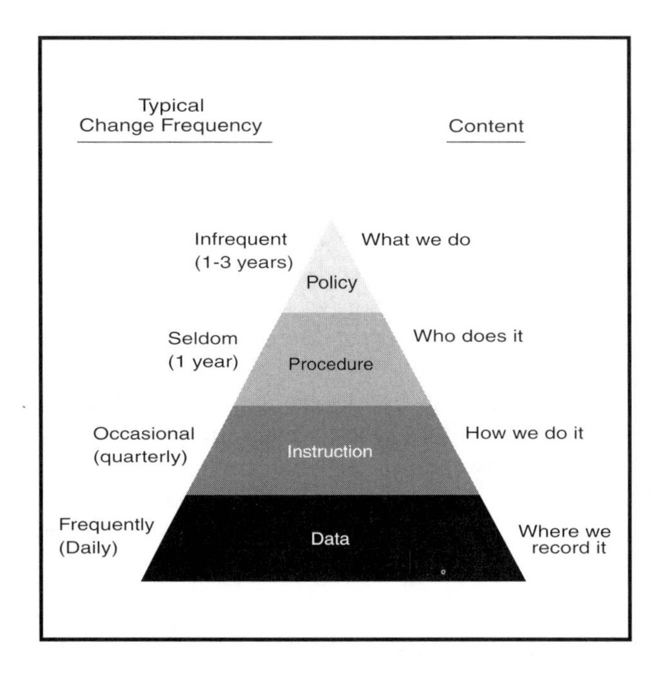

Figure 7.1 The document pyramid.

before you accept the order. Part of that policy says, "We have written procedures for documenting contract review." That is a policy statement; it says what you are going to do, not who does it nor how it is done. Underlying procedures outline who is authorized to accept customer orders, who processes the order, and which hold-points control the release of the order to production, among other items. On the other hand, work instructions associated with the contract review procedure might include a checklist used by order entry personnel to ensure that all pertinent issues are addressed before they accept a contract. Procedures are boundary descriptors because they define critical boundaries in a process. Procedures control departmental assignments and interfaces, as well as detail the movement of data and material. Work instructions, on the other hand, address how a particular piece of a job is done. It is easy to grasp the difference between the two documents if you envision a work instruction being read and followed by a single person and a procedure being read by everyone in a department.

While the distinction between procedure and instruction may seem trivial, it is extremely useful when you are contemplating system maintenance. As the document pyramid shows, separating procedure from instruction reduces the depth and frequency of revisions. Instructions, because they describe the way people do a particular job, have the highest change frequency of all operations documents. If instructions are embedded in procedures, procedures will take on the revision character of instructions (i.e., they will have high revision rates). For example, if the contract checklist is embedded in the contract review procedure, every time the checklist changes, the procedure has to be revised. Naturally, the procedure is distributed to more people than the checklist since it has a larger scope. Boundary functions, such as sales, production, and purchasing, would be interested in who reviews contracts since they need to know where they interface with the review function. They could care less what the contract review clerk's checklist includes. By embedding instructions in procedures, the distribution list now includes people who are either uninterested in the instruction or uninterested in the surrounding procedure. In either case, you wind up distributing revisions to people who have no interest in the change at hand. It is immaterial whether or not you take advantage of computer-driven document systems that claim to make it easy to track relationships among documents. If you do not have discipline within the documents themselves, no amount of outside, precanned discipline will help. You still will inundate people with needless documents. Even if they know little about document systems, they will know that yours is out of control.

If you do well on all three efficiency attributes and have a framework for classifying documents, your document system will be well-behaved. When a new or revised document hits the floor, those who get it will know that it contains vital information. They will read it every time because revisions, although not rare, nevertheless are infrequent and therefore warrant their attention. If you rate poorly on the attributes, people will get frequent document revisions having little or nothing to do with their jobs. Eventually, they stop reading the revisions, and the document system suffocates under its own weight and dies of neglect.

7.12 EXERCISE AND REGULAR CHECKUPS

How do we ensure that our document system and the underlying operational procedures stay healthy? All systems need periodic checkups. People should see a doctor every once in a while; likewise, document systems should be checked. Such a check is called an audit, and it ensures that the underlying documents reflect actual activity and that the system itself is effective. Audits are done in the name of upper management, usually in the name of the premier manager on location. Most standards-based management systems require a formal, documented audit program. We have found that such an audit program is unnecessary as long as the company relies on the documented system to run its business, meaning that the top manager is intimately involved in the system itself. At some point, however, a company becomes too large for top management to have first-hand knowledge about the system's health. Without audits, people begin to hide behind "the system." In those cases, an audit program is an invaluable tool for top management. The key is that the audit function must be real, not merely a prop used to sell your company's program to your customers.

How then do you to tell if your operating policy includes all the elements it should? If you are just starting to document the way things are done in your company, you should consult any number of good management standards such as ISO 9001 or the Food and Drug Administration's GMP (even if you are not in the food and drug business). Those standards contain more or less generic operating system elements that can be used in your business as applicable. The elements can be used to form the core of your first (or rebuilt) operating policy manual. If the policy has an inherent flaw or omits an important element, your documented operating system will generate and communicate the need for a policy review. As long as you listen for it and as long as you understand the fundamental role policy and procedure play in your company, over time you will revise your policy so that it meets all your needs. The management audit and its supporting nonconformance reporting systems provide the necessary feedback loop to change policy.

7.13 THROWING WORK OVER THE WALL

Some people might be eager to point out that the kinds of document systems we have described are at the root of the problem of American business. The classic picture of this problem shows each department working on its part of the process, boxing their work up, and throwing it over the wall to the next department, not worrying about the consequences once that work has been handed off. But wait: Document systems are not the culprit in this situation—poor management practice is. Document systems can only describe and implement management policy. Basic to all systems are (1) embodiment of best practice gathered through the lens of experience and (2) feedback. If management neglects to establish feedback tools, like nonconformance reporting, corrective action systems, and periodic multidiscipline system reviews, the document system will not work even when people, caught up in the system, employ their best intentions. Unmonitored document systems indeed lead to departmental provincialism. But that is a failure of general management oversight. Just because management decides to abrogate its responsibility for maintaining system health because they do not have a clue how document systems work is no reason to abandon systems altogether. The transmutation of a documented operating system into ad hoc, self-empowered work groups is evidence of degenerate management, unschooled in the fundamental skills needed to run a business.

Finally, let us revisit an idea we stated as fact at the beginning of this chapter, that command and control as reflected in documented operating systems does not preclude employee empowerment and flat organizations. It is easy to get lost in a discussion of how to build or rebuild a document system and forget the importance of content and context. A lesson from World War II provides a stark example of the importance of the peaceful coexistence between control and empowerment. In a comparison of German and Russian soldiers on the Eastern front in 1944, it has been commented that the German soldiers were extremely disciplined and considered by many to be more skilled combatants than the Russian fighters. However, German training also emphasized individual initiative and responsibility; that is, common German soldiers were, as part of their strict training, empowered in the sense of the word as it is used in today's fad management movements. This paradoxical emphasis on discipline and initiative allowed German units to survive incredibly desperate battlefield situations on the Eastern front.

Such empowerment was not unique to German soldiers, to the times, or to the armed forces. The point here is that feeble-minded and patterned thinking in American business has led to the wholesale disenfranchisement of an entire fundamental management skill: the comprehension of the role of policy and procedure in running a company. The overriding concern of the purveyors of management fads is to present simple-minded solutions. Seeing through the paradox of the coexistence and codependency of control and empowerment is complex. It does not lend itself to formulaic, faddish solutions but can and must be approached by successful American general managers and their support staff.

Chapter 8

Decision Errors

We have discussed four of the six fundamental skills of general management: profit fixation, technical literacy, change dynamics insight, and the care and feeding of policy and procedure. Common to the practice of all these fundamental skills is good decision making. While it is a cliché that good managers make good decisions with a minimal amount of information, few managers can explain how they think about alternatives and choose one over another. However, a good manager needs both a basic understanding of common decision-making mistakes and knowledge of a few decision-making aids.

8.1 MORE IS NOT NECESSARILY BETTER

Before diving into a brief study of decision theory, we need to address and then set aside our simple-minded attention to the chronic lack of information (or, conversely, our fascination with the brave new world of unlimited information). The core question is: What relationship, if any, is there between the amount of information and good decision making?

One of the tenets of decision theory is that all decision makers are limited by the cognitive capacity of the human mind; it is a fact that we are unable to hold more than a handful of related ideas or hypotheses in our minds at any one time. If we try to collect more than our cognitive limit, our minds must drop something off the list to pick something else up. What that cognitive limit might be is the subject of extensive clinical testing and hypothesizing and is for our purposes unimportant. What is important is that even if you have an astronomical IQ or a photographic memory, the human mind is severely restricted when it forms hypotheses based on perceived relationships among sets of data.

This basic limitation implies that to make good decisions you not only must be aware of the handicap but also find ways to work around it. Mediocre managers who cry about inadequate data are missing the point. They believe that their superior decision-making skills are hostage to their lack of data. Assuming that more data yield better decisions is the most basic fallacy of decision making. Once you reach your cognitive limit, more data usually confuses you unless you have a systematic method for overcoming the natural limit.

We all have experienced our own tendencies to overrate our ability to make better decisions with more data. It usually shows up when you have to make a particularly expensive decision. Say you are deciding whether you should buy a new car. You think you are on the brink of making the big decision, but you want just a little bit more information. When you get it, you find that instead of helping you with your decision it moves you either further away or sideways along the decision tree. Your frustration level increases. You may even decide to revert to the tried and true method of your childhood, "Eenie, meenie, miney, moe." The problem you face in these situations is not with the data; it is with your ability to make sense of it.

To develop tools to help us improve our decisions, it helps to explore the most common blunders. After all, to find a cure you first have to understand the disease. The errors of judgment we will look at are only a few of the many that have been identified by researchers, but we think they represent a good discussion of how well-meaning managers make bad decisions. These errors include groupthink, overestimating control versus chance ("the illusion of control"), framing, sunk costs, representative thinking, availability, and ordering.

8.2 POLITICS TO THE RESCUE: THE ORIGINS OF GROUPTHINK

Much research has gone into how people make decisions, both individually and in groups. The discipline is called *decision theory* mainly because we can only theorize how people go about making decisions. In America, the discipline gets a regular boost from politicians and military planners because they occasionally make tragic decisions that later find their way into the evening news. More important and regardless of the embarrassment factor, national politicians and military leaders are acutely aware that their decisions can save people or condemn them. With this kind of weight on their shoulders, they want to be sure they know how to make the best decisions possible. They also have a long tradition of keeping accurate notes of their discussions and decisions so that future generations can benefit from their wisdom and avoid their mistakes.

Our first example of decision-making bias comes from Irving Janus, who wrote "Victims of Groupthink" in 1972, which was later expanded into the book *Groupthink: Psychological Studies of Policy Decisions and Fiascoes.* Janus uses both successes and failures of momentous political decisions to show how group dynamics can distort rational decision making. His arguments are particularly persuasive because most of us have seen groupthink at work in our companies, but without Janus's exposé it is not apparent what the disease might be. Janus struggled with the paradox that seemingly bright and motivated political leaders occasionally make exceedingly poor and dysfunctional decisions even when copious amounts of good information are available. While he gave many examples of groupthink, his most memorable were his treatment of two crises of the Kennedy administration, the Bay of Pigs fiasco and the Cuban missile crisis. In each case, the decision makers were the same and the emergencies occurred within a year or so of each other, but they had radically different outcomes. One is seen by some as the lowest point of Kennedy's short administration, the other as one of its highest.

Janus defines *groupthink* as "a deterioration of mental efficiency, reality testing, and moral judgment that results from in-group pressures," but any one-line definition does not do justice to groupthink's infectious effects. In its most insidious form, groupthink attaches more importance to group cohesiveness and consensus building than to ensuring that the chosen plan of action is rooted in reality. While Janus examined national policy-forming groups and geopolitical issues, comparable symptoms of groupthink are present in most management groups. The level at which groupthink tendencies are resisted is, to a great extent, a predictor of how well the management group will be able to generate robust alternatives to problems and thereby choose a good solution.

Groupthink manifests itself in several ways. We regularly encounter managements that have an unrealistic belief in their superiority and invincibility, which of course leads to an underappreciation of the competition. This feeling of invincibility comes from the group's overestimation of the personal abilities of the individual group members (i.e., the members are impressed with themselves). That trait in turn affects the company in a variety of ways besides poor decision making. For example, management tends to dismiss their difficulty in reaching consensus when negotiating with unions, customers, vendors, and competitors. It is hard to find common ground if you believe the rest of the world is occupied by dolts and inferiors.

Managers who exhibit groupthink tend to rationalize any disconfirming information that does not fit their concept of the world and their position in it. Disconfirming information in general is discounted because it might force the group to rethink its basic beliefs. The pressure to conform is so strong that group members are reluctant to share any misgivings they have about the course of events. If they did otherwise, the cohesiveness of the group might be endangered. A telltale sign of this pressure to conform is evident when a member brings up an idea that challenges the status quo; the rest of the group will be quick, by vigorous use of accepted thinking, to bring pressure on the member to actively and loudly withdraw the challenge. Although we have not seen it in business management groups, it is said that in severe cases of groupthink certain members appoint themselves as keepers of consensus. They see their role as protector of group decisions by ensuring that disconfirming information is kept away.

How does a group of otherwise bright people succumb to groupthink? Usually they have built close bonds by weathering a particularly difficult time together. During that period of group formation, the actors get a tremendous amount of confirming and reinforcing information. Ultimately, the group emerges from the original crisis or challenge with an overwhelming confidence and camaraderie. Such bonds developed during the course of business battle are especially difficult to challenge, much less abandon.

Eventually, groupthink will destroy the management function of a company. An isolated, overly conceited, and obsessively protective management group is not open to change based on objective evidence. Their decision-making prowess is sterile, and the market will exact its tribute. Meanwhile, the people who depend on those people for guidance are the ones who suffer.

The Bay of Pigs–Cuban missile crisis couplet is a dramatic example of groupthink because it both illustrates groupthink in action and shows that groupthink can be overcome. The story of those two world crises show how one manager, President Kennedy, was able to turn his group around. The Bay of Pigs was a 1961 CIA-directed invasion of Cuba by a brigade of 1,400 Cuban exiles with the audacious objective of fomenting an uprising in the countryside. CIA planners predicted that the revolt would ultimately lead to the overthrow of the Castro-led government. The invasion was in the planning stages when Kennedy became president. His team had three months to consider the CIA plan and accept, reject, or modify it. As a result of their decision to go ahead with the invasion, 1,200 of the 1,400 men were taken prisoner over the two-day battle. And 35 years later, Fidel Castro was still leading the Cuban government.

The official explanation of the fiasco almost exclusively concentrated on the quality of the intelligence information Kennedy had at his disposal. In reality, plenty of preassault information pointed to the almost guaranteed outcome; however, Kennedy's core advisors failed to pursue questioning that would have uncovered the group's faulty decision-making process and illusions. Kennedy and his advisors were affected by groupthink, but how did it initially take hold? Janus says it was the result of a long and hard-fought election campaign in which New Frontier election hype invigorated the incoming president and his close staff. Although some staff members were comparatively young, Kennedy counted many seasoned political sages in his advisory groups. The feeling of invincibility was all around him, and up until that fateful day in 1961, it seemed to be justified. Of course, self-delusion seldom lasts forever. Sooner or later, the system returns to the mean.

In failure, Kennedy was lucky—lucky in that his first prominent presidential decision was such a total and obvious disaster that his inner circle was shaken to their boots. They were forced to question one of their treasured myths: their invincibility. This P-event called into question all their plans, policy, and procedures. If their first geopolitical decision had been successful, it would have reinforced the illusion that their decision processes were healthy when, in fact, they were not. Of course, that decision could not have been the Bay of Pigs but some other decision that—with a little luck—could have turned out right. No amount of good luck could have overcome the groupthink planning of the Bay of Pigs.

The Cuban missile crisis occurred roughly a year after the Bay of Pigs fiasco. On balance, its outcome was favorable for the nation. The same presidential group that bungled the decision on the Bay of Pigs performed admirably during the missile crisis. How did they make the change? While groupthink was not part of Kennedy's vocabulary, he knew intuitively that his process for eliciting advice from his advisors had to change. Many of them after the Bay of Pigs said publicly and privately that they had early misgivings about the assault but felt that such thoughts would have been disruptive rather than constructive. Using the tragic lessons of the Bay of Pigs, Kennedy radically changed his group management approach during the missile crisis. Early in the missile crisis he made his objectives clear to his advisors; once his objectives had been made clear, he also made clear that he expected vigorous and open debate. He changed the way he held meetings and tended to give more time to opposing views than to confirming ones. He changed the entire decision process.

The full-scale assault on groupthink would not have been possible had Kennedy not led it himself. As with the other five fundamental management skills, decision making requires decision makers. A good staff cannot make up for a poor manager. To believe otherwise is wishful thinking.

8.3 GROUPTHINK IN THE BUSINESS WORLD

While we are sure you can think back over your career and find examples of groupthink at work, one press story in particular comes to our minds. In

1995, L. V. Gerstner, the CEO of IBM, reorganized his top management team even though they had had a very profitable 1994. He pointed out that while the profits looked good they were so because of good luck, not good execution. He emphasized that while fat profits are nice, they are sustainable only if they result from a plan. Although he did not use the word "groupthink," he was ensuring that his management team avoided it by seeing the world for what it was, not for what they wanted it to be. The arrogance tag that IBMers have had in the past was not put there by customers and competitors out of envy; there had to be substance to it. Past IBM management must have unwittingly allowed groupthink to take over their company and reinforce a companywide superiority complex of mythical proportions that diminished its ability to accept disconfirming information.

8.4 ILLUSION OF CONTROL

The next decision-making deficiency comes in many flavors. Basically, it causes us to overrate the amount of control we have over the outcome of a planned event. Two of the most familiar flavors are what we call *one-sided chance* and *denial of the mean*, which are different sides of the same coin.

At first glance, it seems odd that a book that emphasizes the role of planning in successful business would conversely emphasize that people tend to overemphasize the amount of control they have over their lives or their businesses. But these are not mutually exclusive positions. The fact that in most cases probability is a much better gauge of outcome than the skill of the player is something we have to inject forcibly into our planning and decision making. Otherwise, we develop dysfunctional plans that omit a basic truth of the world we live in.

Assigning our triumphs to our personal strengths and skill while relegating our failures to bad luck or chance are the most common manifestations of illusion of control. Few of us consciously assign our triumphs to luck and our failures to personal weaknesses. In other words, many of us accept a probabilistic world only when we want to rationalize our shortcomings, thus the term *one-sided chance*. We like to believe that we affect a good outcome more than we actually do.

An example of someone taking advantage of our tendency to engage in one-sided chance errors is state-run lotteries. When Texas lottery officials first introduced a six-number lottery, they did not offer a feature common and available in other states with more mature lotteries: machine-generated random numbers. Ostensibly, that was because the system and equipment were not ready to issue so-called "quick-pick" numbers. On the contrary, it could have been. After all, the contractor running the Texas lottery ran lotteries in states offering a quick-pick feature. Rather, lottery officials knew that if they offered this feature early in the lottery's introduction, they would diminish the illusion of personal skill and control over picking winning numbers. Television ads for the Texas lottery overwhelmingly emphasized method and skill over chance when addressing success.

In business, assigning good financial outcomes to your skill and poor financial outcomes to bad luck (e.g., a bad market) is delusional. Good or bad financial outcomes are a consequence of both skill and luck. The trick is determining how much of each element might be responsible for a particular outcome. Suppose your good fortune was almost entirely because of a good market; if you had better skills, you would have made an even bigger profit. If you are inclined to assign your success to your skill, you will avoid making any changes to your approach to business. "If it ain't broken, don't fix it." When the market does turn against you, you will be in much worse shape than had you taken advantage of a good market to improve your operations. Instead, you spent precious time congratulating yourself on the good job you were doing. Conversely, suppose your poor results during good times were entirely because of poor decision making, but you assign them to bad luck. The outcome is similar to thinking erroneously that good fortune comes from great management, but it comes much quicker because you are unable to make money even in good times.

Denial of the mean is similar to one-sided chance errors. It is hard to accept that most processes have a mean and that the best predictor of an outcome is the mean. Several examples will help illustrate. Take personnel evaluations. Supervisors historically get frustrated with employees because they seem to be inconsistent. A particularly difficult employee, on occasion, will perform admirably; an exemplar employee will have a bad day. In the case of the difficult employee, just about the time a supervisor believes that the employee's troubles are over, the employee again disappoints. There is little mystery what has happened. The employee's behavior clusters around an average. A good manager will have a systematic way of assessing that average performance level. A poor manager will continue to be bewildered with this "inconsistent" performance, seeing a pattern of improvement or diminishing performance where there is none. The manager essentially refuses to believe that a system (the person's behavior) reverts to its average.

A second example of denial of the mean is typical company financial forecasting. The ability of a company to generate profit is generally stable over short periods, say a year or two, unless it is going through some radical refinancing, restructuring, market explosion, or similar upheaval. However, many large companies require their divisions to generate monthly rolling 12-month predictions of profit and cash flow. Because of the technology of spreadsheets, these reports are dutifully generated. They usually show upward trends in all the important criteria even if the historical average is flat. If they show deteriorating conditions, the deterioration in the future months will always be forecast to be less severe than the historical average. That is a denial of the mean. The best forecast is your current mean performance.

You can also find examples of denial of the existence of the mean in articles written by people who should know better. An article in the membership journal of the American Society for Quality Control explored the different ways winners of the Malcolm Baldrige National Quality Award enhanced intracompany communication. In the article, the director of employee communications at Motorola was quoted as saying that he depended on face-to-face meetings between employees and managers to get the word out about company policy and goals. While he admitted that reliance on the supervisors made good communications a function of the skill of these managers, he said that had not been a problem with Motorola. Besides smacking of groupthink ("We're different, we're smart, we're invincible"), his statement totally ignored statistical reality. Motorola had over 132,000 employees at the time. There was no way anyone could build a system of supervisory training and performance evaluation that eliminated the bell curve in a company that size. The manager was implying that Motorola had some mysterious way to take a population of, say, 500,000 or more potential employees and filter them so that Motorola could identify and avoid the poorest performers. That is the only way that average supervisor performance at Motorola could exceed that of the average corporation. It did not, and it could not. In such a large worker population, you cannot avoid average performance no matter how smart you think you are. Making decisions based on denying the existence of the mean is not rational.

Last, take a typical sport like the stock market or baseball. Invariably the financial pages will say that some nag stock has finally come around and will be the next Microsoft. Or, in the middle of the baseball season, the sports pages will be full of stories that a certain struggling pitcher has turned the corner and is back to his old form. In either story, reality is much less dramatic.

Take the pitcher's situation. On average, major league pitchers develop the same way: building strength in the minors, getting occasional assignments in the big leagues, joining a major league rotation or bullpen, reaching peak performance level, staying there a few years, and, ultimately, deteriorating. That is what the averages say. What they do not say is that on any particular day few pitchers perform at their average. Additionally, few of us are taught that, because of cognitive limitations, perfectly random events—like having a good day or a bad day—*appear* to occur in lumps (i.e., people tend to see patterns where mathematical testing shows none exist). This cognitive tendency may be vital to our individual creative and survival skills, but it is an illusion. In sports, they call it a hot streak or a cold streak, depending on our perception of which way the streak is going. However, sports streaks do not exist in the probabilistic, real world, only in our minds.

A pitcher who seems to be coming out of a slump is really reverting to mean performance. One who seems to be diving into a "new" slump is merely exhibiting the fact that, on any particular day or during any particular period, average performance may be hard to see. Over the longer term, it is inevitable.

So if most performances are not average performances, why pay attention to the average? Because it is the best predictor of a future outcome. In business, if you think you have a hot hand and bet on it staying hot, on average you will pull up short. If you compound your decision-making delusion with a belief that your hot hand is a result of skill, then you are doubling a bad bet. Your luck may continue, but eventually it runs out. Take the task of choosing a chief executive officer. To some extent, board members in search of the next CEO choose the wrong candidate when they say they look only at "results." If they fail to explore how much of the candidate's previous success was the result of a hot hand and how much of it was skill, they may find themselves in the equivalent position of signing a pitcher who starts reverting to his mean performance the day he shows up at spring training.

8.5 A NICE FRAME AFFECTS THE PICTURE

The decision traps we have just described are easy to overcome compared to avoiding framing effects. It has been shown that, depending on how possible outcomes of a decision are worded, your preferences for certain outcomes can be influenced. That effect is most evident when the outcomes of certain alternatives are described as losses or gains. People seem prone to pay more to avoid a loss than gamble on a gain, even when the outcomes are identical. A common example used to illustrate that effect is of an acquaintance calling you up at 3:00 in the morning. If he says, "Listen, I just figured out how you can make $5,000," you likely would tell him to call you at a more decent hour. However, if he told you that he had found a way to keep you from *losing* $5,000, you probably would stay on the line a little longer. But why should you act differently depending on the proposition? The monetary outcome of either piece of advice is the same, assuming your friend has a sure bet (i.e., a riskless decision). In either case, you would be $5,000 richer than you would be otherwise. In the first case, you would have $5,000 more tomorrow than you have today. In the second case, you would have the $5,000 tomorrow that you otherwise would have lost.

This decision error phenomenon is exploited by sellers of canned management fads. Take the selling of ISO 9000 quality system registrations in the 1990s. We can predict that most chief executives will respond more favorably to a pitch that such registrations would prevent them from losing market share than that such registrations would help them gain market share.

Put yourself in the position of a typical CEO of the day. Assume for the moment that the chances that this fad can deliver either of its promises—avoiding a loss or securing a gain—are the same and that the size of market you lose or gain will be the same. In such a case, the monetary outcome is equal. In either case, if you do not choose to buy into the fad, you will be poorer for it by the same magnitude. For example, say your division sells $30 million a year. If you do nothing in the first case, your sales could drop to $25 million. You now are $5 million poorer. In the second case, if you do nothing, your sales remain at $30 million, but you could have increased them to $35 million. In either case, if you do nothing, you will have $5 million less than you would have otherwise.

The presumed outcomes are identical, but the first frame is much more persuasive. As an aside, someone might claim that the second choice, in which you increase market share, is twice as good as the first because you would both avoid losing $5 million and increase sales $5 million, thereby netting $10 million. The salesperson might try to sell you that line, but the two claims are mutually exclusive. The only way you could see yourself $10 million richer is if a third outcome is postulated: that registration both avoids losing market share *and* increases market share. That appeal is used infrequently because it mixes the emotional appeal of loss avoidance with the less fulfilling appeal of obtaining a gain. In any case, the predominate advertising appeal of the ISO 9000 fad of the 1990s was fear of losing market share.

Another example that takes advantage of framing effects comes from retail sales. A large wine merchant we know charges customers for using credit cards. However, the charge is framed as a cash discount, not a credit card surcharge. Why? Because imposing a surcharge would imply a loss, which customers would avoid by not using their credit cards. But shoppers look at the retail, nondiscounted price as their frame of reference. If they choose to use cash and take advantage of the "cash discount," that is a gain that they are happy with. If they choose to use their credit cards, the merchant ensures that they never come face to face with their loss.

These examples show that the effects of framing are subtle and difficult to overcome. They are also unrelated to semantics. Researchers have constructed ingenious two-choice tests that, because of framing effects, cause most people to swap their selection when the frame is changed. When the subjects are told that they swapped their decisions from a positive outcome to a negative outcome solely because the frame changed and are given a chance to repair their inconsistencies, they more often than not choose to remain inconsistent. The decision techniques we present later may, when taken as a whole, help limit the natural bias most people have to pay more to avoid a loss than to capture an equal-sized gain. However, the best long-term solution is to insist that the potential outcomes of every proposal you see or develop yourself are framed consistently and preferably as gains.

8.6 SUNK COSTS MIGHT AS WELL BE AT THE BOTTOM OF THE OCEAN

A cousin to framing is valuing sunk costs, something most managers learned in finance class but which has application to all decision making. Valuing sunk costs when making decisions is of particular interest politically in these days of budget restraints. During the debate on continued funding of the Super Collider project, many people pointed out the vast sums of money that had already been spent as justification for continued spending. The question at hand was not whether the money already spent were spent prudently, the question was whether the additional money that needed to be spent would deliver value in excess of the proposed amounts.

Cost overruns almost always occur because people honor sunk costs and disregard the lessons sunk costs teach. How else can we explain $350-million public works projects that end up costing billions? Each time additional funds are requested, two arguments are presented. First, "only" an additional couple of hundred million is needed, and, second, we have already spent hundreds of millions, which we have to "protect." Two other points are rarely brought up. First, the hundreds of millions already spent over the budget have bought us, at most, the knowledge that we do not know how much this beast ultimately is going to cost. Second, at this point in the project's development, we might have enough hard evidence to predict that the benefit of the completed project will never exceed the incremental cost.

Suppose the taxpayers are asked to spend $350 million (on top of the already $800 million spent) on some military weapons system. The contractor admits that changes in military technology and changes in potential threats have reduced the effectiveness of the weapons system to, at most, $300 million. (The number was calculated by accepting the fact that the weapons system's performance improvements cannot be fully utilized in the new battlefield scenarios; therefore, the supposed improvements are valueless.) The fact that we are being asked to pay $350 million more on a system that will deliver $300 million of value is lost in the argument about "protecting" the original $800 million.

As we said, we honor sunk costs in many other ways. For instance, suppose you are ready to buy a new car. Your current one is getting on in years and is becoming unreliable. You decide what make and model you want and how much you want to spend. While you are making your decision, your old car has transmission problems, and you shell out $700 to fix it. If you decide to postpone purchase of a new car in order to get your $700 "back," you are honoring sunk costs. You cannot get your $700 back. The only value your $700 has is confirmation of your decision to buy a new car because your old car is, indeed, unreliable. If you insist on honoring sunk costs, you will never get a new car because you will always be trying to get the last dollar you spent out of your old heap.

Another example: Suppose you spend $100 on some expensive tickets to an entertainment event next Saturday. On Friday, that special someone you have been trying to attract shows some interest in doing something with you on Saturday. You turn the object of your affection down in favor of your original plans, even though you would rather go for entertainment of a different kind. You are honoring sunk costs. The money has already been spent, and nothing you do will get it back. You are better off looking at that $100 as the price of doing something you want to do (being with that someone special) rather than the price of doing something you would rather not do (going to the event).

Sunk costs can also come in the back door. We know of a Fortune 500 company that held a sizable receivable from a management consulting firm. Receivables are sunk costs if you cannot recover them. Instead of recovering cash or writing it off, the company agreed to hire the consulting firm for a morale-improvement program to "recover" the sunk costs. Of course, the company never viewed the receivable as sunk. Instead, they spent additional time and money making their staff participate in an ill-conceived and generally unwanted program, all because the company's management honored a sunk cost.

Honoring sunk costs in business is different from honoring sunk costs in your personal life, and that may be where the rub is. In business, sunk costs are not honored because the underlying assumption is that the business is ongoing and has an infinite life. People do not have infinite lives and are very aware that they all face the same end. If you are 60 years old, you are less likely to start a new career than if you are 35. As a 60-year-old, you must honor sunk costs. You have spent a lifetime building a skill and a career. Disregarding the effort you have put into your life because it is a sunk cost is irrelevant and irreverent. Personal decisions have to recognize that life is finite; economic business decisions do not.

It bears repeating that the only thing that can be brought forward into the future when contemplating sunk costs is not how much you have spent on something so far, but what you have learned from the money you have spent. Knowledge is transportable into the future, sunk costs are not.

8.7 SO, HOW DO YOU THINK HE WILL REACT?

A good salesman we know, when asked how he thinks a particular customer will react to a blind proposal, always reminds us that he has no idea. By doing so, he is studiously avoiding representative thinking. *Representative thinking* is a cousin of denial of the mean. When you deny the mean, you see trends where none exist. You believe that you can forecast subsequent events of stable systems by relying on short-term, near-time events rather than on mean performance. When you use representative thinking, you overemphasize the importance of one piece of information you possess over others you do not possess.

In the case of our salesman, he refuses to assign any importance to clues he may have concerning the willingness of a potential customer to buy. Why? Because he knows that he hardly has all the appropriate information to make such a judgment. He might judge that there is a 5% chance that his mark will respond to his proposal, based on his hit rate for potential customers in that class. By doing so, he is honoring the mean. But to believe that he has a better or worse chance with a particular customer because of other factors is an illusion.

For example, suppose he notices that a potential buyer has a big title and works for a big corporation. If he is not careful, our salesman can slip into representative thinking and conjure up all sorts of reasons his hit rate might be lower. He thinks, "Big Titles working for Big Companies have assistants. Assistants can be overprotective of access to their bosses. Therefore, the chances that the blind proposal will get by the assistant is less than average." He has just invented a scenario that has less probability of being true than if he were to guess that his target was no longer with the company and therefore never receives the proposal.

The fact that we can imagine many possible scenarios of how an event might play out has no effect on the fact that the real world generates more possibilities than we can hope to conjure. Sure, by working for a bigger company, the target has a higher chance of having an assistant than if he worked for a smaller company and had a lower-sounding title. But what of it? That may *increase* the probability that our salesman might score a hit based on other scenarios he has not thought of. The assistant may be looking for some way to impress the boss and decides to respond to the proposal. The world is too complicated to allow our feeble mind to assess outcomes based on the number of scenarios it can generate. You can almost guarantee that your concept of how an individual will react or how a situation will turn out will be wrong unless, of course, you have access to objective statistics. Then you will be wrong only half the time.

A more thought-provoking and personal example of representative thinking is the typical reaction many of us have when we see an ill-kept man on the street corner and immediately think he is a potential criminal. While it is true that a majority of petty criminals do not dress particularly well, the vast majority of poorly dressed people are *not* criminals and therefore do not warrant your distrust. If you are having trouble understanding that point, remember that serial killers are, for the most part, normal looking. Both Ken Bianchi, the California Hillside Strangler, and Ted Bundy, the University of Florida Co-ed Killer, were well-dressed, articulate serial killers. Does that mean that women should avoid well-dressed, articulate young men? We do not think so.

8.8 I SAW IT ON TELEVISION

Availability is perhaps the most frustrating decision-making error because it is the easiest to avoid. It seems the more we know about a certain situation, the more we overestimate the probability of it happening. Decision theorists say these events are available to us through some personal experience.

For example, customers asked to estimate the percentage of defects in your product would overestimate the rate. Why? Because in the absence of hard data, the customers would remember the problems they have had with your product more accurately than they would be able to recall how much product they buy from you.

Availability is also at work at a much larger and more expensive scale. After the Exxon Valdez spill in Alaska, additional safeguards were mandated in the U.S. tanker fleet, including double-hulled tankers. That Valdez-type accidents represent only a small fraction of the overall risks of oil tanker operations and that money might be more wisely spent on reducing more significant and predominate risks were lost in the passion of the moment. The Valdez accident was available—the fact that it happened distorted the more important discussion about overall tanker safety.

Continuing with the oil industry, look at something called "wellbore design." Wellbore design is the arrangement of below-ground equipment used to produce a well. For the most part, the equipment is nonproprietary, off-the-shelf stuff. Strangely, wellbore design varies considerably in the same field. It is not uncommon for well costs to vary as much as 50%, depending on who designs and drills the well. Some of the difference reflects rational differences in risk assessments, but most of it is the consequence of the distortion of availability in decision making. The wellbore design policy of different oil companies tends to reflect not the long-term experience of the industry but the near-term experience of the individual oil companies. If a company has had a recent rash of high-profile blowouts, it will change its policy toward design and operations in response to the recently available experience. If it has had a rash of collapsed pipe, it will beef up associated designs. Wellbore design varies across the industry because base failure rates are ignored in favor of incremental failure rates that are overly affected by episodic, high-profile, available events.

8.9 SO, WHAT DO YOU WANT TO EAT?

When you go out to eat, how do you decide what you want? You scan the menu and the first or second item that strikes your fancy, you order. If you could observe yourself, you would find that more often than not you choose the last dish that strikes your fancy, not the first. Later, you look over at another table and notice that someone is eating your favorite dish and you missed it on the menu. You just became a victim of ordering.

Ordering in decision theory refers to the tendency we have to choose the first alternative that appears to satisfy our needs. If we continue our search, we do so by comparing the subsequent items to the last one that appealed to us. If we find another appealing one, it replaces our definition of acceptable. We continue the search against the last appealing alternative. People in real estate know this tendency well, which is why they usually start out showing you houses you utterly detest and save the best for last.

Ordering indicates two things. First, we stop our search short of exhausting the list. Second, if we continue a search, we resort to anchoring our decision with the last acceptable choice. To some extent, we have to accept ordering as a natural limit of our resources. At some point, we have to say that we are finished generating alternatives and have to make a choice. Additionally, any search for a good answer has to include some scheme for ranking alternatives.

The problem with ordering is that there are better ways to make decisions, though not necessarily when it comes to choosing a meal at a restaurant. (The restaurant example is a nice introduction, but it is a weak analogy for business decisions. Few business problems come with a ready-made menu of choices.) In business and to some extent in your personal life, if you fail to separate the generation of alternatives from selection, you will suffer from poorer decisions than if you did otherwise. The dysfunctional approach of combining the search with the decision means that you are held hostage by the order in which the alternatives are presented. That gives an inordinate amount of weight to your first acceptable (though not necessarily best) alternative. There is more evidence that ideas generated later in your search will be more robust and fulfilling, but poor decision techniques eliminate them before you can generate them.

8.10 EPILOGUE

This brief discussion of the decision process was just that, a discussion. We avoided rigorous definitions and thorough theoretical discussions in favor of a leisurely walk through the discipline of rational decision making. It was a good setup Chapter 9, which introduces a handful of decision tools meant to help you avoid typical decision errors. Decision theory as a discipline is, of course, much more elaborate than it appears here. If you want to really delve into this meaty subject in more detail, the appendix contains a list of books that explore the subject. We highly recommend a trip to the library or bookstore.

Chapter 9

Decision Tools

Identifying the root of a problem is not the same as solving it. In Chapter 8, we visited the more common decision errors. Some problems suggest their own solutions, while others seem unavoidable. In this chapter, we introduce a few tools you can use to improve the quality of your decisions. You will not necessarily use all these tools at any one time. Some of them, like decision trees, you may never come back to.

The central idea of all these tools is that to improve decisions you have to find ways of avoiding the bad ones that lead to disaster. Some people will see that as planning for failure, which they will say ensures failure. We do not deny that success depends, to a certain extent, on psychological factors. Success may indeed hinge on going into uncertain situations with every intention and belief that you will succeed, but turning that thought into a cliché does not increase its authority. Planning for failure is the best way to remove the downside, which allows people to be more secure, not less, in their assessment of success. Certain warrior tribes went into battle only after they had made arrangements for their funerals. They had no intention of dying. Ensuring that the downside was protected meant that they were more confident of the upside.

9.1 PAPER AND PENCIL VERSUS SPECULATION

In Chapter 8, we discussed business meetings and why most meetings are universally dismissed as colossal wastes of time. We touched on the importance of keeping a record of all meetings if we are going to have a chance of increasing their value. More generally, you will not be able to improve your business decisions at all if you do not figure out how to use paper and pencil in all decision situations.

It is easy to trivialize the importance of a simple piece of paper and a pencil, but bear with us for a moment. Take out the notes from your last meeting. Go through them briefly and circle one or two important points. Do not spend more than a minute on this exercise. Once you have marked what you think was important, find someone else who was in that same meeting. Ask that person to review his or her notes and tell you what the salient issues were. More than likely, you will find that the other person does not have any notes, cannot find them, cannot make any sense of them, or has an entirely different impression about what was discussed and decided. The last problem—disagreeing on what was discussed and decided—is not really a problem. At least both of you can recall your impressions from contemporaneous notes. The unsolvable problem is that in the absence of meeting notes, any value the meeting might have had at the time is completely lost. You might as well have gone to lunch than had the meeting.

The written word provides the sustenance of civilization. The next time you see a television news clip of a working meeting of the U.S. president or a congressional committee, notice the nondescript staff people in the background whispering to each other while earnestly clutching sheaves of paper. The next time you channel-surf, stop at C-SPAN coverage of someone giving testimony; notice that the witness refers to notes of a conversation held two years ago. The next time you follow the progress of a lawsuit, notice how testimony based on contemporaneous notes carries more weight than someone's unassisted recollection.

On the other hand, have you ever gone to a business meeting to discuss an important question and no one took any notes, much less issued minutes after the meeting? How often in those circumstances did the meeting degenerate into a forum for unstructured speculation and one-upmanship? Ever been frustrated that after such a meeting all the principles had different ideas on what was decided? Have you ever wished you could go back and recollect the events of that meeting but know that you never will be able to? This typical portrayal illustrates an irritating trend by today's managers: Many ignore the tools of their trade or feel uncomfortable displaying those tools, all to the detriment of their work.

A manager who does not know how to use a piece of paper and a pencil is like a construction carpenter who does not know how to swing a hammer. Neither can do the job. Such dysfunctional behavior is surprisingly widespread and seems to be driven by an influential pop culture stereotype that colors management's view of the world. The image of the American power manager includes tailored suits for her; starched, monogrammed shirts for him; and for both, a very thin, leather portfolio holding a few clean sheets of paper and an expensive pen for signing those important documents. The updated image has them using a computer, so weightless and small that it fits into that ever-present thin, leather portfolio. Never mind that the streets of Manhattan, Los Angeles, Chicago, Houston, and a thousand other business centers are actually populated by men and women in rumpled clothes, lugging 5.5-pound laptop computers in over-the-shoulder bags, loaded with candy bars, computer disks, operating manuals, and three sets of keys. They dream of the day they can off-load their white-collar flotsam and jetsam and join the power managers. Disregard the occasional glimpses of the captains of American business at work. They almost universally are overweight, bleary eyed, and not particularly well-dressed. They also lug around huge amounts of paper wherever they go.

Mind you, managers should rely on their staffs to do staff work. Likewise, there is nothing wrong with wearing nice threads. However, the power look—the right clothes and the right portfolio—is a high-maintenance masquerade that can be pulled off by very few people. Most of us will never work at a job where we have a staff fully dedicated to supporting our individual, personal effort, even if we reach our goal of running a company, division, or department. If we have a staff, they most likely will have their own work to do in addition to assignments related to our work. We find ourselves in positions that demand we dress the part of power managers, but we actually accept a different position. We might walk the street looking good, but back at the office, at home, and in the car, stacks and stacks of paper await our attention.

9.2 THE FINE ART OF NOTE-TAKING

Let us return to that business meeting in which the discussion degenerated into a marathon contest of speculation and put a pad of paper and a pencil in front of every participant. At least now they have a prop; if they get bored, they can draw caricatures of other people in the room. Say you are in the room. What would you do with your paper and pencil? First, when the meeting is over, you should have some sort of record of the ideas that struck your fancy and maybe a note or two to yourself concerning what your next action will be. Second, during the meeting, you should be using that paper and pencil to march yourself through some of the decision tools we will discuss later. Ideally, those decision tools are displayed on a flip chart for everyone to see. (If your peers are not up to your speed yet, you will not be using that flip chart, but that is a different problem. You still will need your paper and pencil to apply the tools for your own benefit.)

If you are not used to making meeting notes, you may wonder what you should write down. This is the fun, artistic part of your job. Like art, there are a few useful rules, but you still get to react to what you hear and see, and you get to create your own landscape. Rule 1 is that your notes be honest and clear to you. There are no rules on how to organize, what format to use, and what your notes should look like. You will develop your own style. We like to jot down revealing phrases or comments preceded by the name of the actor. If we have a personal reaction or thought concerning the point just expressed, that response shows up in brackets. That helps us get closure on ideas that boiled up earlier but were inappropriate to mention at the time.

In essence, personal notes are a history meant for your eyes. They are your means of communicating to yourself over space and time. The value of a personal record becomes evident when you go back to them weeks later. As you read your notes, you will notice that they include important items that you never could have recollected without the help of a contemporaneous record. For example, you might come across a note concerning Sam's awareness of a particular problem. At the time you scribbled it down, it did not mean much, but when you read it later, you may be able to put it in some useful context. "I wasn't sure Sam knew about the sales situation back then, but here it is in my notes. Yesterday he told me he didn't know about it until after the meeting." Sam was not lying yesterday; he actually believes he was not privy to the sales problem at the time of the meeting. His honest recollection differs from the facts. Without notes, his mind is incapable of keeping temporal track of critical events. In the absence of a written chronology, the human mind is more liable to invent history than to recall it completely.

Habitual note takers know how unreliable memory is and that notes do not need to be exhaustive to have value. In fact, personal notes are a work of art, not a labor of engineering. If you get wrapped up in note taking, believing it to be a structured and demanding process, you will miss sharp turns in the discussion and shifts in tone and attitude. A short record using key phrases and comments will enable you to recollect much more of the meeting. Such notes serve to jolt your memory later, not replace it entirely. Sometimes meetings are so intense that you will be unable to make any notes during them. If that is the case, take a few minutes afterward to reduce your recollections to paper. Better then than later. Better later than never. Our experience is that notes taken within 30 minutes or so of an event, although different from those taken contemporaneously, are a good recollection of business events. On the other hand, you might as well skip notes altogether if you cannot get them done before the end of the day.

We think we can say without hyperbole that the invention of paper and pencil was as necessary to civilization as the taming of fire and domestication of wild grasses. You are fortunate, through education and opportunity, to be able to apply the power of paper and pencil every day. That power allows you to overcome your natural cognitive limitations, multiplies your intellect, and serves as the essence of leverage in your work. It allows you to order alternatives and thereby is the basic instrument of good decision making. Do not abandon it just because popular culture says that people with paper and pencils are nerdy pencil-necks. The most successful, richest self-made people in America are nerdy pencil-necks. There has to be a connection.

9.3 THINK ALTERNATIVES

You may still question our homage to pencil and paper. But as we lay out the fundamental elements of good decision making, it will become evident that good decisions cannot be made intuitively. Now that we have put a pencil in your hand and a piece of paper in front of you, we will talk alternatives.

The first step in good decision making is defining alternatives. Every problem that needs action is a dilemma, a decision that requires choosing from alternatives that, on the surface, seem equivalent yet may not be. The first alternative in any list of alternatives is the status quo. The question "Why change?" is cardinal in the change framework presented earlier in this book because it forces decision makers to recognize that the status quo must be fully repudiated if we are to move on to other alternatives.

After ignoring the cardinal question, the next routine and predictable mistake decision makers make is concocting a flimsy list of alternatives. This is the first place a piece of paper and a pencil come in handy. Since most people's cognitive limit is five items (plus or minus two), you cannot keep all the alternatives in front of you, so to speak, unless they are physically in front of you. So whip out a piece of paper and start listing as many alternative solutions as you can. Do not think about how good or bad a particular alternative is; just jot them down. There will be plenty of time to figure out their attributes and assign them ratings.

Suppose you are unhappy with your current job and decide to look at your alternatives. If you limit your list of alternatives to either staying with your present company or finding another job, it is too easy to decide to look for another job. You have just limited the quality of your decision. Your alternatives are limited by your hasty move to the next problem statement, "What other jobs are available?" You need to stay at the "Why change?" question much, much longer. Why are you unhappy with your current job? Depending on the answer to that question, you might come up with this list of alternatives:

- Do nothing: Stay with my current job.
- Drop out and go on the road with a bar band.
- Ask for a transfer to a different department.
- Go back to school.
- Change professions.

- Move to another part of the country.
- Take up an avocation.
- Seek counseling.
- Ask spouse to go back to work.
- Ask for a raise.
- Get a pet.
- Have an affair.

The more alternatives you can generate without prejudice, the better chance you have for clearly defining the problem and finding a good solution.

9.3.1 Dominant Alternatives

By writing down your list of alternatives, you avoid putting a halo on any one alternative and prevent, as best you can, anchoring yourself to an alternative generated early in the process. We cannot overemphasize the importance of generating alternatives without concern about how realistic, practical, or sane they might be. Ignoring the attributes of an alternative lets you include some outlandish possibilities that, while initially outrageous, may lead you to think of related but more practical possibilities.

Much decision making does not have to go much further than listing alternatives, because one choice becomes the obvious dominate alternative. In our sample list of apparently equally attractive alternatives, the dominate one might be to seek professional counseling.

Even if you do not spot a dominate alternative, you might be able to narrow the possibilities down to a few. Suppose you narrow the list down to two choices: stay with my current job or ask for a departmental transfer. Rather than getting overexercised about the decision, write the competing alternatives across the top of a sheet of paper. Then write the pros and cons beneath each. In our example, your list might look something like Table 9.1.

Table 9.1
Judging Alternatives

Stay	*Ask for Transfer*
Enjoy my co-workers	New friends
Cannot stand my boss	New boss
Good money	Less money
Have a track record	Low seniority
Topped out	New potential for growth

A simple list makes it clear not only what you consider important about your decision but what risks are inherent in your decision. It also is one way to look availability right in the eye. Remember from Chapter 8 that availability is the decision error that makes us believe outcomes available to our memories are much more likely to occur than they really are. For instance, most people believe that it is more likely that they will be murdered than is the case because they hear evening news reports on the high crime rate. Lists like the one about the pros and cons of leaving our current job always make the unknown, future state look riskier than something familiar, even though the amount of risk may be equal. For instance, you might like your co-workers and therefore believe that staying in your old job will ensure a continuing relationship, but, in doing so, you are not considering the probability that they themselves might quit. You cannot stand your boss and figure that the current situation will last forever, but you underestimate the chances that your boss may transfer before you.

Listing the pros and cons of a couple of major alternatives helps you focus on which attributes are more important than others. Suppose, in our example, when you start the process you think the "potential for growth" attribute overwhelms all the others. After listing the other attributes, you may find a balance between attributes that you would not have recognized if all you had done was think about the problem instead of writing it down. If after looking at the list you still cannot decide, you can either flip a coin or get more rigorous.

9.3.2 Mini-Max: The Quickest Way to an Obvious Answer

But before we get more rigorous, we need to touch on a line of thinking called the *mini-max solution,* that is, *mini*mizing your *max*imum regret. Suppose you decide you are going to ask for a transfer. Before you act on your decision, ask yourself, "What is the worst that can happen if I go ahead versus what is the worst that can happen if I stay? Which decision has the potential for encompassing my biggest regret? Is there any way to minimize my maximum regret?" Suppose you would regret it more to ask for a transfer and not get it than you would just to stay in your old job. Why might that be? You could be concerned that your friendly co-workers might treat you differently back at your old job once the cat was out of the bag. Since your relationship with them is important, you may not want to jeopardize it. You could also be concerned how your already poor relations with your boss might be affected by an unsuccessful transfer request. While you still may move toward that transfer, mini-max thinking makes you take time to gather more information about the probabilities of getting that transfer. You might also do more planning for your reentry into your old job if the odds go against you. The mini-max solution allows you one last chance to check the validity of your decision. It asks you, the decision maker, to develop a worst-case scenario and see if you can survive without taking a fatal hit.

Managers give mini-max solutions too little attention. They tend to believe that alternatives have only one possible outcome. Unfortunately, those predicted outcomes generally are weighted toward the manager's preconceived ideas about the desirability of the alternative. For example, new-product introduction plans are notorious for overestimating the good things that can happen while ignoring the downside. A prelaunch mini-max review will at least show where the company is most vulnerable; while not necessarily killing a proposed action, it might be able to reduce the impact of a poor outcome.

9.3.3 Scoring Alternatives

If you cannot narrow your alternatives down to one or two and make a choice by simple inspection, you need to use more sophisticated decision-making tools. If you are struggling with a real dilemma, a cursory review of the list of alternatives may show that several are equally attractive for different reasons. One will seem particularly good for one reason, while another could be the hands-down favorite for an entirely different reason. You need a quantitative method to rate each alternative against its competing alternatives. The alternative whose assumed outcome has the best rating is your choice.

Let us try a new example. Say you have decided to open a distribution facility in New Jersey. To make the example simple, suppose you have narrowed the choices to two alternatives. Location A is in a heavy industrial area near a major interstate interchange. Location B is in a light industrial area in a New Jersey suburb. You list the following decision attributes:

- Location relative to truck transportation routes;
- Location relative to trained work force;
- Lease cost;
- Length of lease;
- Proximity to local vendors;
- Physical condition;
- Alternative lease opportunities in area.

This list of attributes is no more sophisticated than the one we generated for choosing whether or not to ask for a departmental transfer, except that it is stated without any reference to the alternative we are rating. Your next step is to compare each of the two alternatives and give them a rating in each of these areas. The ratings you give are arbitrary, but we will use a scale of 1 (bad) to 5 (excellent).

- *Location relative to truck transportation routes.* The best location would be near a major interchange, because that would keep your turn-around time down. The worst would be in a rural area served by a two-lane blacktop 30 miles from the nearest interstate. With that in mind, you give location A a rating of 5. Location B is better than being 20 miles outside of Wink, Texas, so it does not deserve a 1; you give it a rating of 2.
- *Location relative to trained work force.* The best location would offer college graduates at prison inmate wages. The worst would give you prison inmates at college graduate prices. You give location A a 2; location B a 4.
- *Lease cost.* This is an easy one. The best location would give you rural West Virginia rents in New Jersey. The worst would give you Tokyo rents in New Jersey. In this case, location A and B are about equal, so you give them both a rating of 3.
- *Length of lease.* This is an interesting one in your case. You want a short-term lease because you are thinking about minimizing your maximum regret in this attribute. Since this is a new location, it may not work out, so you want to be able to pull up stakes easily. The best alternative would give you a month-to-month lease; the worse would lock you in for a couple of years. You will not entertain anything longer than two years. Location A wants a one-year lease, which is not bad for prime industrial lease property in New Jersey, but no prize. You give it a rating of 2. Location B will give you month-to-month after six months. You give it a rating of 4.

You get the picture. Decide how the best and worst alternatives would rate for the attribute, even if such choices do not exist in the real world. Then, rate each alternative against this yardstick. Give each alternative the same rating if they are essentially the same. You should, however, still give them a rating consistent with your best-worst case. Later, we will show you how to rank alternatives and weight your scores, so keeping true to your rating system is essential.

At this point you have a piece of paper with the information contained in Table 9.2.

The column "Ideal Location" illustrates what the score would be for the perfect location. Disregarding that nonexistent option and all things being equal (meaning each attribute is equally important to you), you would choose location A since it is nearly 10% better than location B and is 66% of the unattainable ideal site. But all things are never equal. Some attributes are obviously more important than others, but how much more? How can we quantify a qualitative issue? The answer is that we have to weight each attribute against the others.

Table 9.2
Distribution Center Decision

Attribute	*Location A*	*Location B*	*Ideal Location*
Transportation	5	2	5
Work force	2	4	5
Lease cost	3	3	5
Lease length	2	4	5
Proximity to vendors	5	2	5
Physical condition	2	4	5
Alternative locations	4	2	5
Total raw score	23	21	35

We do that by first ranking the attributes from most important to least important, which makes rating them easier. Suppose you rank your distribution facility attributes in this order:

1. Length of lease;
2. Condition;
3. Cost;
4. Transportation;
5. Local vendors;
6. Alternative locations;
7. Work force.

Rankings are easier to make if you first pick the most important, then the least important, then work your way to the middle. At this point, you have not said how much more important item 1 is over item 2, just that it is relatively more important.

The next step is matching numbers to your personal feelings of relativity. Again, the number system you use is immaterial, but suppose we stick with a rating scale of 1 to 5. (If you had a longer list of attributes, you might choose a rating system of 1 to 10.) Next, assign a rating of 5 to your most important attribute and a rating of 1 to the least. By that measure, "length of lease" gets a 5.0 and "work force" gets a 1.0. Then, look at the second most important attribute, condition. How much less important is it than length of lease? This is where your personal feelings are supposed to get quantified, not ignored. Based on your bad experience with poorly maintained property and the importance of avoiding facility problems during startup, you place a pretty high weight on the condition attribute. You give it a 4.5, indicating that it is not as important as the length of lease, but nearly so. You continue rating each subsequent attribute until you get to the bottom. If there is too large a jump between the next-to-the-last and the last attribute, you may want to go back and adjust your ratings. Suppose you come out with the rating listed in Table 9.3.

Table 9.3
Distribution Center Decision—Attributing Rating

Rank	*Attribute*	*Rating*
1	Length of lease	5.0
2	Condition	4.5
3	Cost	4.0
4	Transportation	3.0
5	Local vendors	2.5
6	Alternative locations	2.0
7	Work force	1.0

Now, you want to normalize the ratings, that is, make them add up to 1.0 rather than 22.0 (see Table 9.4). You do this by adding the ratings and then dividing each attribute rating by the total (e.g., 5 / 22 = 0.23 for the first attribute). The reason for doing this soon will be apparent. (We promise.)

Table 9.4
Distribution Center Decision—Normalized Rating

Rank	*Attribute*	*Rating*	*Normalized*
1	Length of lease	5.0	0.23
2	Condition	4.5	0.20
3	Cost	4.0	0.18
4	Transportation	3.0	0.14
5	Local vendors	2.5	0.11
6	Alternative locations	2.0	0.09
7	Work force	1.0	0.05
Total		22.0	1.00

So far, you have quantified your feelings about the importance of each attribute to your final decision. Now, you will take each weighting factor and multiply it by the original, unweighted rating you gave each alternative. For example, transportation's rating at location A would be 5 • 0.14 = 0.70. (See Table 9.5.)

Table 9.5
Distribution Center Decision

Weight	*Attribute*	*Location A*	*Location B*	*Ideal Location*
0.14	Transportation	0.70	0.28	0.70
0.05	Work force	0.10	0.20	0.25
0.18	Lease cost	0.54	0.54	0.90
0.23	Lease length	0.36	0.92	1.15
0.11	Proximity to vendors	0.55	0.22	0.55
0.20	Physical condition	0.40	0.80	1.00
0.09	Alternative locations	0.36	0.18	0.45
Total weighted score		3.01	3.14	5.00

When the attributes are weighted, the new score indicates that you should change your mind and choose location B, which is about 4% "better" over location A.

But your job is not finished. Remember our caution about taking models at face value. You cannot. The implication that this decision model forecasts the best option is false. Like any tool, it is more important to know what it cannot do than what it can do. The questions you have to ask include: What caused the preference to change from location A to location B? Is a 4% difference important, or are we essentially talking coin flip here? If it is a coin flip, what is the next important attribute we have not listed or do not want to—perhaps the distance from our house to the new facility? Are the weights right? Are the ratings and rankings right? How much of a change in our ratings would cause the decision to flip again? The power of this decision tool is not that it makes the decision for you, but that it illustrates subjective considerations and helps you avoid or at least uncover the decision errors and biases presented in Chapter 8.

9.3.4 The Complication of Uncertainty

The decision matrix ignores one unpleasant complication. Decisions are problematic because they involve uncertainty, and the decision matrix implies that the outcome of your decision is certain. It tackles the question of how to value outcomes, not how to handle the uncertainty. It helps order your preferences logically and systematically; however, it asks you to state the outcome of a future decision as a certainty when, in fact, you cannot be certain. In the case of choosing the location for your next distribution warehouse, we rated the two alternatives by assuming that the predicted outcome was certain. For example, when we rated each alternative on the attribute of physical condition, we said that location B was in better shape than location A. (Location B had a rating of 4; A had a rating of 2.) In fact, we predicted that B was 100% better than A. We do not know that for a fact; that is just an expression of our relative feelings about the physical condition of the two locations. We may have inspected the properties and even hired a building inspector. Nevertheless, we are quantifying our belief of how each stacks up against the other. For all we know, location B might have a latent electrical problem, which burns the place down after we move in. A decision matrix papers over the role of uncertainty in decision making, but it is a good compromise between just musing aloud and building a model that includes the estimates of fire experts concerning the probability of fires caused by hidden electrical faults in buildings less than five years old.

There is, however, one more decision tool you might want to use to model your probabilistic forecast of uncertain events. While knowing the potential effects of hidden defects may not be a cost-effective use of your time, you might want to model the effect uncertainty has on your decision to sign a long- or short-term lease.

9.4 DECISION TREES

Decision trees are the classical tool of technical staffs dedicated to solving sophisticated decision questions. These tools lend themselves to modeling probabilistic outcomes that have monetary impacts, but they also can be applied to nonmonetary outcomes as long as you can quantify your preference for one outcome over another. Decision trees supposedly are used by industries that regularly make large capital investments that hinge on future, uncertain events. Natural resource development, like mineral prospecting and hydroelectric projects, are paradigm investments for which decision trees are used to illustrate the impact of such uncertain events as future political climate, product cost, and expenses. Companies that play those games know that their assessments of the likely outcomes of such attributes have a considerable impact on their decision to invest. Like all models, this tool does not give a "go" or "no-go" answer. Rather, it shows how *estimates* of uncertainty might affect a decision. It is a way to test the sensitivity of your decision to the amount of uncertainty in the future outcome of that decision.

Here, we will omit all the preliminary work necessary for a solid foundation in the use of decision trees because that is not what we are after. Probably the most complicated tree you will ever draw will have four branches at most. Many times, you will not even "solve" the decision tree but instead use it to illustrate the structure of a problem. What every decision maker needs is a grasp of the concepts—a simple example to come back to when there is a real-life problem to solve. (If you need a better understanding of this tool so you can communicate with experts who use it for a living, refer to the appendix for an appropriate reference.)

Let us return to the question of which location you should lease for your distribution warehouse. Say the landlord at your preferred site surprises you and makes you a double-sided offer. You can either lease the property for six months for $10,000 per month with a $1,000 escalator at each six-month anniversary for a maximum of two years, or you can sign a two-year lease at a flat $10,000 per month. Which one you choose depends on your estimate of the chance the warehouse operation will be successful.

Figure 9.1 shows the basic decision. The two-year fixed lease will cost you $240,000. The six-month lease with the escalator will cost $276,000 over two years. If you remained for at least two years, you would be $36,000 richer if you took the long-term lease. However, suppose you think there is a high probability that your distribution center will fail. Under such circumstances, the potential $36,000 savings has to be balanced against your possible losses connected with getting out of a two-year lease. You need to know the effect of this uncertainty on your decision.

Figure 9.2 takes you through the first uncertain outcome. Regardless of which lease you take, the question remains, "What are the chances that the warehouse will last past six months?" To simplify the analysis, we will assume that if it lasts six months, it will last two years. You assign the probability that it will close in six months as 90%; the chance that it will last two years as the balance, or 10%. You have no quantitative information to help you set those probabilities; they merely reflect your confidence in your answer. How do you get to a 90% failure rate? First, without any information about the process under consideration, the chances of any two-outcome uncertainty is 50%, so that is where you start. Suppose you have had a bad track record opening new locations. That tends to make you pessimistic about the potential for the warehouse, but instead of avoiding all expansion possibilities, you want to ensure that all the decisions associated with your opening reflect that pessimism. Besides, you tell yourself, a 90% failure rate is not

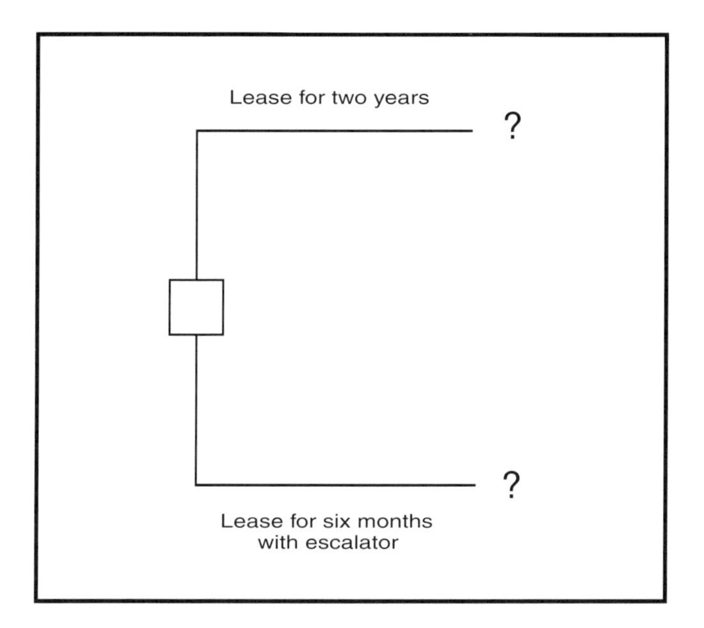

Figure 9.1 Basic leasing decision.

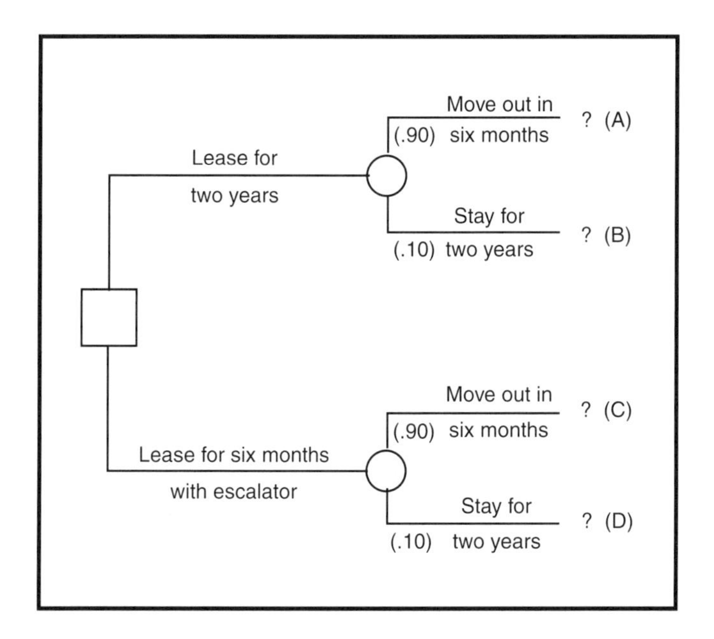

Figure 9.2 Leasing decision.

uncommon for new businesses of any kind, and each new location is essentially a new business.

Each branch of the decision tree in Figure 9.2 has a monetary outcome. To calculate the topmost outcome, labeled (A) in Figure 9.2, you have to model one more uncertainty. Instead of suffering the complete loss of your two-year lease if you move out prematurely, you might be able to negotiate out of it or sublet. Suppose you think that you have a 90% chance of getting away for the equivalent of one year's lease payment, or $120,000. Your confidence comes from past experiences of negotiating out of leases gone bad. That means you figure you have a 10% chance of getting stuck for the complete $240,000. That could happen if the economy heads south in six months, making it impossible to sublet. Figure 9.3 shows how this additional uncertainty is added to the tree. Outcomes B, C, and D shown in Figure 9.2 are straightforward. If you stayed for two years under the long-term contract, you would pay $240,000. If you moved out after six months under the short-term contract, you would pay $60,000; while if you stayed two years, you would pay $276,000. But none of these outcomes is certain. To calculate the value of a branch on a decision tree, multiply the outcome's value by the probability that the outcome will occur. The cost of the decision to lease for two years is $142,800; the cost of the decision to accept the short-term lease is $81,600 (Figure 9.3 shows the calculations). It is obvious that you should go with the short-term lease.

Or is the decision so obvious? The power of a decision tree is not that it can predict which alternative is more attractive, but that it shows the effect of your *assumed* probabilities on your decision. We always like to play with the probabilities to determine at what point a decision will flip. We then ask ourselves, "How close are we to that flip point, and how can we get more information to improve our risk assessment?"

Assume you feel confident of your chances of negotiating out of the two-year lease (90%), but you are not so confident of your assessment that you will be closing the warehouse in six months. A simple algebraic solution shows that the flip point for uncertainty is 33%. That is, if you were to assess your chances of moving out at 33% or less, then going with the two-year lease is better than going with the short-term lease. Figure 9.4 illustrates that flip point. Unless you can come up with convincing data that erase your initial pessimism about the project and since the model is relatively insensitive to the possibility of moving out early, you probably would stick with your original decision to sign a short-term lease, even though it would cost more if the distribution center is successful.

Decision trees inject reality into business decisions that have become overwhelmed by overly optimistic forecasts. Expansion projects like our distribution warehouse example tend to build momentum and are pushed along by groupthink and the honoring of sunk costs. Opening a branch

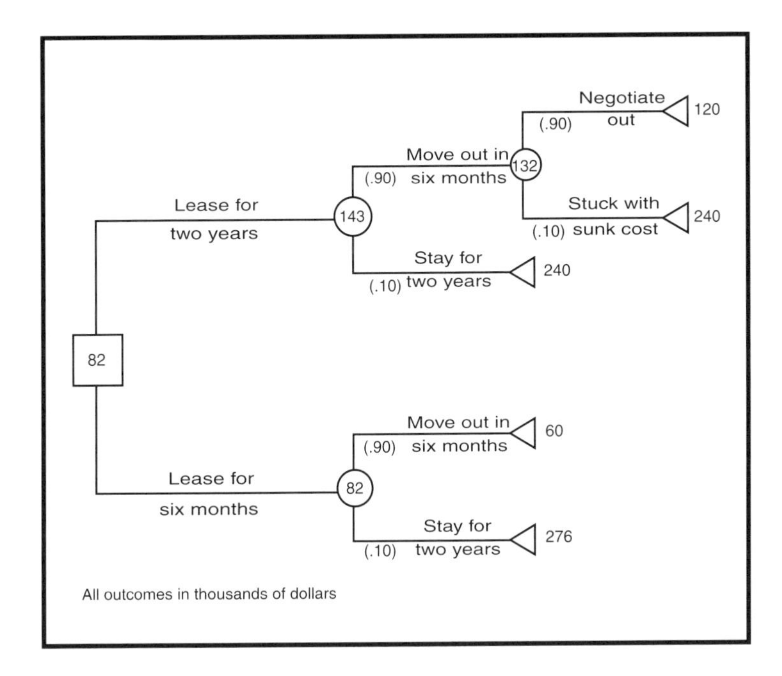

Figure 9.3 Leasing decision.

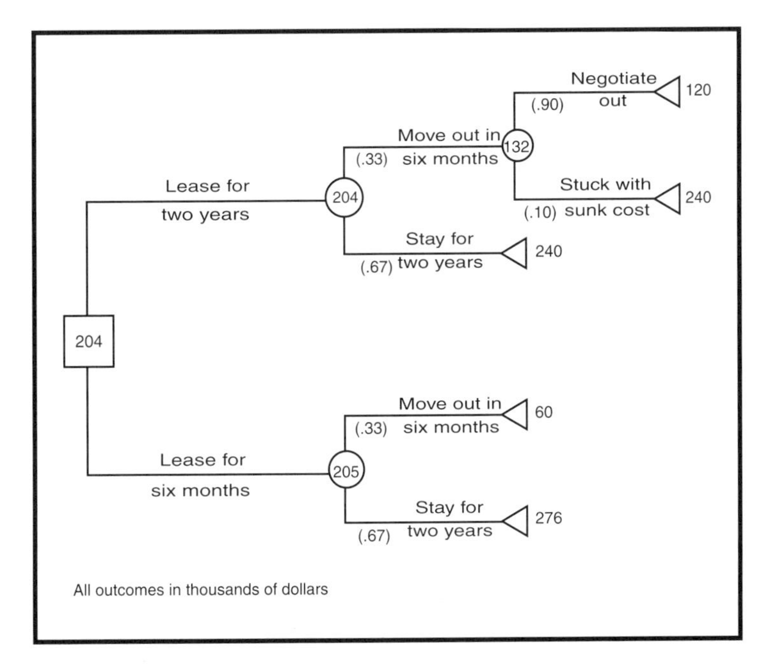

Figure 9.4 Leasing decision flip point.

location is a typical decision situation in which enthusiasm on the ground is crucial if you are going to improve the odds for success. That enthusiasm has to be tempered when you are looking at limiting the risks associated with the commitment to move ahead.

Again, no decision tool should be used exclusively. For instance, in this example you need to apply mini-max thinking if you decide that your initial pessimism is unwarranted. Suppose your sales force comes in with fantastic forecasts and a survey saying that the new location will be a hit. You have a chance to make an easy $36,000 by taking the longer lease. However, you have to think about minimizing your maximum regret. You may rather forego the possibility of getting $36,000 in order to minimize your maximum regret of being stuck with an empty warehouse and a commitment of $240,000, even if the possibility of that outcome now seems remote. Mini-max solutions are even more appropriate when the decision you face is fairly unique for you. If you were opening new locations every two weeks, then you would have plenty of opportunities to take advantage of the odds in your favor. If opening a new location is a rare event, you want to avoid gambler's ruin—the distinct possibility that even if the long-term odds are in your favor, your early play will be peppered with a string of bad luck.

9.5 DECISIONS ARE OPTIONAL

The last decision tool we want to explore is option thinking. In the example in Section 9.4, mini-max thinking could help you avoid a worst-case situation. Another way of looking at the extra $36,000 you would spend on the short-term lease is to classify it as the cost of an option. An option is an investment that allows you to make an investment later (or, conversely, to avoid making an investment). Conventional options are financial instruments that allow the owner of the option to buy or sell a stock at a given price before a certain future date. Options come in other forms: You can negotiate an option that allows you to buy a certain piece of land at a certain price in the future. Buyer of options are not forced to buy or sell; they simply have the option to do so for a specified period. In our warehouse example, you could look at the additional $36,000 you would pay with the short-term lease as the price of an option that allows you to avoid an empty warehouse and a negative cash flow of $240,000. It is an option because you may not need it. You may find yourself staying in the facility for two-years under the short-term lease and not need the protection of that option.

Translating investment decisions into options is coming into its own, but option structures are more difficult to state and understand than other decision models, like decision trees, that imply a fixed answer. However, with a little ingenuity, most decisions can be coached in the language of options. Take the familiar decision to invest in a new machine. Suppose you can buy one large machine with a certain fixed output or three smaller machines whose total production equals the larger machine. Buying and operating the larger machine is less expensive than buying the three smaller machines. If you believe you can load the larger machine, then the choice is obvious. However, the decision is not without uncertainty. How certain are you that you will be able to load the large machine? In that case, the additional, incremental cost of buying and operating three smaller machines rather than one large one can be seen as the price of an option. The option allows you to adjust your production to meet actual demand, thus avoiding having a less-than-100%-loaded monster machine. Your option also includes the opportunity, but not the obligation, to sell the excess productive capacity if your forecast demand is less than the three small machines. Because of the uncertainty in your forecast and your ability to see the decision as an option (rather than a loss), you may indeed decide to go with the more expensive alternative of three small machines. The transformation of a decision into the terms of an option gives you another perspective on decision making and provides another avenue to avoid the decision errors discussed in Chapter 8.

9.6 INTO THE BREACH

The most useful idea in this whole chapter is that you will go far if all you do is use a pencil and paper to arrange and cultivate your thoughts, eliminate alternatives, search for obvious dominate strategies, and avoid common decision errors. If you use the tools described here to limit your downside risk, you will feel more comfortable with making difficult decisions, thereby moving forward more quickly and more confidently because you know you did your best thinking. In the end, decision making is a numbers game. The more decisions you make, the better the chance you will make some good ones, especially if you ensure that the bad ones do not kill you.

Chapter 10

Duality

Art imitates life. In the 1993 movie *The Fugitive,* Tommy Lee Jones plays a U.S. marshal charged with bringing in convicted wife-killer Dr. Richard Kimball, who had escaped with another inmate. In one scene, the posse corners the other escapee, who has taken a hostage. The marshal distracts the bad guy long enough to get in a few lucky shots. Later, when an associate tells Jones's character that he cannot believe the marshal did not try to negotiate first, Jones says in a slow whisper, "I...never...negotiate." That scene, along with other telling ones, builds the marshal's character into a straight shooter, so to speak.

As the movie builds toward its climax, the marshal displays fatherly attention to the well-being of his team, admits a few mistakes to his fellow posse members, and even shows conditional sympathy toward Dr. Kimball. Rather than detracting from the business side of the marshal's character, this soft side actually enhances the audience's appreciation of the marshal as a genuine person. Scriptwriters call this "character balance," and it is as necessary in the real world as it is in good storytelling. Successful managers from first-line supervisors to CEOs must master their dual role of leader and comrade.

10.1 DUALITY: A NECESSITY

The sixth fundamental skill of manageering is the ability to project the duality of leader and comrade. All of us project different personas as a means to communicate more effectively. Growing up, we discovered that we had to interact differently with our sisters and brothers, our parents, our grandparents, our teachers, our friends. In some ways, those relationships represented different worlds, in some ways they were the same. Many times, such relationships overlap, sometimes they never can. For example, while your son may someday become your friend and get to enjoy your "friendly" face, he will never become your father and therefore will never see your childlike face.

You can see this adaptability developing in four-year-olds as they generate different appeals to different people in their lives. They act one way with their mothers, another with their fathers, and another with their aunts. They are so brazen about this facility that they change their tune mid-sentence when they figure out they are using the wrong persona. As we get older, we become more tactful but no less adaptable. Eventually, a person is able to shift from one persona to another without the awareness required of a four-year-old trying to learn the ropes. Somewhere in our mid 20s, most of us begin to encounter fewer new authority figures and competitors. The rate at which new social situations requiring a new persona appear begins to decrease. Like our capacity for learning a new language, our capacity to handle such situations also decreases.

Handling the duality between being a boss and being a comrade is one of the last big persona development tasks you have as an adult. It is not a trick or a manipulation but a requirement of successful communication and leadership. But by the time you need to develop duality, your skills for persona development are not what they used to be. What we do in this chapter is give you a few tools to increase your awareness of duality and help reawaken skills long forgotten.

10.2 A CHANGE PROJECT FOR EVERY MANAGER

Duality is such a difficult capacity to develop that few escape the learning process without some scars. Many of us are familiar with inept managers who were artificially harsh when a situation called for understanding, only to be overtly but disingenuously kind when they tried to recover. We also have seen or heard about bosses who avoided confrontation like the plague and those who could not live without it. All these people were struggling unsuccessfully with their duality.

To these managers, the sensitivity training fads of the 1990s were a disaster for various reasons. Being tough was out, being kind was in. Managers were encouraged to be more sensitive; employees to be more assertive. In an attempt to reeducate old-style managers, those classified as too abrupt, too blunt, or too harsh were sent to charm school to smooth a few rough edges, while employees enrolled in assertiveness training classes. The implication was that changing management style was not unlike learning to get along with some of your in-laws. It might be distasteful at times, but in the interest of the family it is better to find a way to tolerate them. The faddish reeducation schools claimed that the new management style was more professional, more acceptable, and more successful than the "old" way.

The content of those management reeducation programs is immaterial because their objectives are inappropriate and unattainable. Except for the short-term damage done to participants doomed to fail and the waste of time and opportunity, those faddish schools cannot change anything. By this point, you know why: Changing something as personal and inseparable as management style is the most complex change initiative there is, subject to all the problems mentioned in the previous chapters. Whether we are trying to make managers more "people-oriented" or more "business-oriented," the results are similar. Back on the job a few weeks or at the most a few months, the "bad" managers begin to backslide. It is agonizing to their subordinates as well. They can tell the managers are trying hard to become "better" managers, only to be frustrated by the difficulty of trying to be someone they are not. Managers classified as "too tough" find that no one believes their metamorphosis is real. Those managers told they were "too undemanding" never become comfortable with asserting themselves.

When the principles of change dynamics are applied, it becomes obvious that a company program for management reeducation is universally unworkable in America. First, we have trouble getting past the first standard question "Why change?" Very few managers, even those who are convinced that their style is the root of their company's problems, will be able to answer "Why change?" unequivocally. Remember, that question is asked during the unfreezing stage and revisited continually, consciously or unconsciously, throughout the change process. Second, defining and justifying a company, division, or department change objective in terms of a personal change objective is problematic: What is the cost? Where is the profit? How long will it take? How will we know when we are finished? Third, look at a typical force-field analysis of a personal change initiative (see Chapter 5). What are you going to show as forces that encourage and that oppose change? How is an unbecoming management style a serious threat to survival? What is the source of disconfirming information? Can it be collaborated with external information? Are the external sources of disconfirming information strong, consistent, and insistent? Fourth and last, how are you going to measure change? Typically, management-style changeovers are justified on the basis of obvious-by-inspection, supposedly needing little project-specific economic justification. But as we have said before, in business, if you cannot measure it, it does not exist.

We are not saying that changing management style is an inappropriate *change* objective. We are saying that it is an inappropriate *business* project. Few personal change projects are appropriate as company- or departmentwide projects. How about undertaking a physical fitness change objective? An abstinence initiative? Why not put together a program to enhance management and employee spirituality? While some companies have embarked on those types of change initiatives, such programs are outside the definition of a business change program. They are inappropriate because they cannot pass the tests of a business initiative. Of course, what employees or managers do on their own time is not the company's business. Whether it is profitable to encourage private behavior by providing facilities for such activities is a very different question left as an exercise.

10.3 DUALITY REQUIRES CONSISTENCY

If duality is a fundamental management skill independent of style, how do we develop it? First, we have to have a better handle on the nature of duality; second, we have to find tools to help us reinforce that nature. Duality is a skill that good managers have for appearing simultaneously as a leader and as a comrade. Therefore, duality is a personal skill, apparent both in group settings and one on one. If we are to reinforce duality, we have to have a strategy for both situations.

Duality was listed as the last of the six fundamental skills because it is enhanced by the other fundamental skills: profit fixation, technical literacy, change dynamics insight, policy and procedure comprehension, and decision theory knowledge. Having a profit focus helps improve your consistency, thereby enhancing your leadership presence. Organizational psychologists have various competing theories about what makes a leader, but most agree that people universally respond to consistency. That is natural. Without consistency, followers have trouble forming pictures (i.e., gestalts) of reality consistent with their leader's. Furthermore, followers seldom get closure if their leader, with little preparation or forewarning, continually changes focus. Understanding how people change and the role that awareness plays in that process improves the personal side of your duality. Additionally, while people like to like their bosses, they are more impressed by bosses who are technically competent than those who are good storytellers. The techniques of good decision theory add to your cachet as a competent group leader, and understanding your role in policy and procedure formation ensures that you intervene at appropriate places and times.

The bulk of this book has been taken up with ways to develop a balanced portfolio of business management skills. Those skills lend themselves to objective measurement and can be improved using practiced methods. Duality depends on developing personal relationships in a business environment and is less conducive to formulaic solutions. Nevertheless, a few ideas might help improve this last and most enigmatic fundamental management skill.

10.4 SHARED DECISION MAKING

Bosses always complain that being at the top is lonely. They say they have no one inside the company to talk to, and when they do confide, they feel they pay for it. They find that more is made of their inquiries and questions. Instead of getting a handle on a problem, they find themselves inadvertently feeding the rumor mill. In retrospect, this syndrome reflects a different kind of inconsistency on a manager's part. In addition to being inconsistent when it comes to goals, objectives, policy, and procedure, managers can be inconsistent with the way they include or exclude people from their decisions. The balance between being seen as a leader and being seen as a comrade is influenced by the topics you reserve for yourself and those you share with others. If you are consistent about what topics are appropriate for each forum, you find that people respect and understand your duality.

In the 1979 film *All That Jazz*, the protagonist gets up every morning and prepares himself for the day in the same way: Before going out, he looks in the mirror, puts on a smile, and says, "Show time!" Likewise, every time you are at work as a manager, some part of you says, "Show time!" You know that someone will be watching and listening, trying to extract as much meaning as possible from your mood, your demeanor, your little asides. Too many managers want to forget that they, more than anyone else, must think before they speak, verify that what they have said was fully understood, and recover if they have misspoken. Someone once said the mark of a professional is doing your best even when you do not feel like it, and many times managers do not particularly feel like being "on". They want to let their hair down; they want to have a few friends on the inside.

People want their leaders to be human, too, but their allowance is not as great as it is for themselves. What a leader such as yourself chooses to discuss freely must follow a pattern, and that pattern reinforces your duality. If one time you muse out loud about the company's financial condition and the next you treat it as a state secret, you will be seen for what you appear to be: an untrustworthy confidant. If you encourage people to tell you what they do not like about the company and then do not do anything about their complaints, you will be seen as what you appear to be: disingenuous. If you ask them their opinion about a particularly difficult decision that affects them and then make a decision that totally ignores their input, you will be seen as what you appear to be: conniving. Never mind that you are none of these. Your actions speak louder than words.

10.5 THE INCLUSION TREE

The prerogatives of leadership include responsibility for a decision when all is said and done. Your constant dilemma is that you cannot make robust decisions without getting information from people around you, but as soon as you engage someone you have to ensure closure. The dilemma is solved if you have a strategy for inclusion. We call that strategy an inclusion tree, a reference to the decision tree presented in Chapter 9. You must ask yourself the following questions when you are deciding whom to include or exclude:

- *Does the decision lend itself to analysis, or are we talking about preference?* It makes no sense to include other people's opinions if the final decision rests on your preference. Be brutally honest with your answer to this question. If you consistently lie to yourself and say, "Yes, this decision lends itself to analysis, and I need other people's input," and then do what you want anyway, the people around you will learn to withhold information routinely. Then, when you really need their input, you will not get it. On the other hand, if you include people in your decision only when you really need objective input, they will be more prone to share their knowledge.
- *If the decision lends itself to analysis, will the answer be enhanced if I include other people's input?* The glib answer is, "You always should include other people's input. You're a cretin if you don't." We disagree. In some cases, you do not have time to include other points of view. In some cases, the analysis is simple and direct. In other cases, security of the decision requires limited distribution (i.e., notification on a need-to-know basis). If it is more important to protect the decision than it is to improve its quality, make the decision yourself or with limited input. For example, some complex change initiatives require the agreement of many people and alignment of forces for change (e.g., introduction of a new product line). On the other hand, some complex change initiatives require the agreement of few people and require a high level of security (e.g., reducing headcount during a severe recession).
- *If I need other people's input, whom do I include to ensure a good decision?* While this question is stated as inclusive, deciding whom to exclude is actually harder. Most managers find it easy to include and routinely find that they have included too many nonessential people. In any case, the reasons for inclusion or exclusion must be shared with those involved. If people think they should be included and are not, they react badly if they are not told why. Suppose you exclude someone because you do not trust that person to be objective, even though he or she may have a valuable point of view. You have got to be honest.

Let the person present a case and participate on a conditional basis, if you must, but do not make a big mystery out of it. Intrigue is good for Sherlock Holmes, but it has no place when you are concerned with the quality of a decision. On the other hand, you may have to include diverse, closed-minded opinions if implementation of the proposed decision requires consensus. Again, it makes no sense to pretend that differences do not exist. They do, and you must address them up front.

10.6 LIMITING INCLUSION

That last point brings up a couple of issues that must be emphasized. The first is the tendency to include more people in a decision than is necessary or prudent. The second is the sticky problem of excluding people who believe they should have been included.

The faddish trend in management education has been to extol the virtue of embracing as many diverse opinions and ideas in the search for a solution to a problem. The inclusion tree leans against that wind. In our pursuit of improving the quality of decisions by widening the information net, we have ignored the importance of context in shaping an inclusion strategy. Taking an example from our files, the president of a company that designs and installs architectural accessories decided to embark on a team-based, problem-solving program. With the president's approval, the outside consultant encouraged middle management to brainstorm the problems that needed attention, which they decided were salary administration and fringe benefits. They drew Pareto diagrams, assembled lists of information, and assigned work goals. The wheels fell off the program for several reasons, but a climb through the inclusion tree would have prevented the disaster in the first place. Salary administration topics usually require limited participation to ensure the security of the process. Otherwise, expectations rise, personal agendas are aired, and forces opposing change build prematurely and fatally. Besides that, the president would have answered "No!" to the second inclusion-tree question ("If the decision lends itself to analysis, will the answer be enhanced if I include other people's input?") had it been asked. Therefore, the context of the decision determines inclusion rules, not some philosophy of the more the merrier.

Note how the inclusion tree actually works in building duality. It does not teach managers how to "act." It does not try to form new personas. It provides an umbrella of consistency during the time managers are developing their own, free-standing personas. Eventually, managers who use their long misplaced persona-building skills will get a grip on their own duality. To flesh out the true-life example in the preceding paragraph, you should know that the president was new to her job, having advanced out of the independent sales force. The firm's professionals decided to test the president's grasp of her duality, and what better way than to pick on a central management prerogative: salary administration. The president, having just left a role that was primarily that of comrade, had little time to bolster her leadership presence. If she had had the advantage of the inclusion tree, she at least could have avoided the challenge while borrowing time to learn from the situation.

The second point that the last inclusion rule brings up usually is avoided in polite management discussions: the role of conflict and employment termination in improvement programs. Supposedly, companies are enlightened places where people spend stress-free lives improving themselves and their companies. Termination is the mark of a weak manager unable to help a troubled employee and a symptom of Neanderthal employment practices. Not necessarily. Conflict is a natural state when two or more people hold dear opposing opinions that cannot be resolved without one side or the other (or both) changing an essential principle of their understanding of how things work. Conflict is the essence of any change initiative. Whether those conflicts can be resolved in the time available determines whether or not one side or the other can remain in the group. The importance of the required change and of the relative strengths of the conflicting ideas is left on the manager's desk to decide the next move. Some changes cannot be left to eventual resolution, and termination is the only solution.

10.7 DUALITY: A STATE OF MIND

What is important when you are considering the inclusion tree is not the absolute issue at hand but its relationship to other issues. Deciding when to include or exclude other people in your decisions reflects both personal choice and the company's "culture." The important ingredients are that management be consistent, that shared decisions have the same relative importance, and that those decisions left to the prerogative of leadership are likewise related. If you treat the assignment of weekend duty as one that requires limited input but the assignment of company cars as a group decision, it will be hard for people to understand how inclusion works.

As you grow accustomed to recognizing instances of duality, you will see that it takes many forms. A rock-and-roll story shows how good managers of all persuasions recognize duality's importance. In the late 1980s, ex-Beatle Paul McCartney produced a documentary of his American tour. At that point in his career, he also produced his own music, so he had a million and one things to look after and about the same number of problems to solve. A telling point in the film comes when the interviewer is talking with the members of McCartney's band. One band member says that when Paul shows up at rehearsals and sound checks he never talks about the business side of the tour. All he shares is his music and his excitement about the tour.

McCartney's duality was a refreshing and uncommon trait, and for the music side of the business a critical skill if band members were to concentrate on their part of the tour's success. McCartney knew this, and both avoided including them in issues for which they had nothing to add (finance and front-office operations) and instead nurtured their awareness of issues on which they had substantial impact (music). The band also recognized the situation and respected McCartney in some ways more for the management skills he brought to the organization than for his awesome musical abilities.

Earlier in this book we mentioned in passing that all types of management styles can be successful. Dysfunctional management is not a problem of style but one of consistency, predictability, and trueness—trueness meaning being true to your natural character. In the end, that sincerity is the source of your individual duality, not some tool like the inclusion tree.

Chapter 11

Fundamentals Versus Fads

The principle contention of this book is that technologists have a ready-made professional disposition to be good general managers but that they abandon their valuable insight in favor of other, less valuable management perspectives. Now that we have covered the six fundamentals of manageering, fundamentals drawn from the experiences and training common to technologically trained individuals, we are ready to test the power of those fundamentals. What better way to do that than to reexamine popular management fads using the lens of manageering fundamentals? As has been our standard practice, we will take a few unexpected, contrarian turns while on this course, so fasten your seat belts.

11.1 THE FUNDAMENTALS OF FADS

You probably have read about popular business fads failing: TQM, empowerment, ISO 9000, reengineering, or any one of a half-dozen other prominent programs. Maybe you have heard about the company that won a national quality award but filed for bankruptcy a few years later. Perhaps you yourself have suffered through some program, complete with speeches, slogans, teams, and new but obscure language, only to see it sputter to a quiet stop on some abandoned road like so many before it. Yet even in the midst of countless examples of failure, the business press still spins stories of success and triumph around those very same programs. The inspiring fables usually include CEOs pulling their companies from the brink of disaster using "innovative" techniques. Suddenly, these people become the newest business heroes and their programs the latest business fads.

Despite the occasional and overexposed success story, the vast majority (we would say 60 to 80%) of business fads fizzle. How is it then that New (Business) Wave gurus still find followers? What are the rules of the game? Is it a game worth joining? What separates success from failure? And what is it about business fads that makes them more appealing than a sincere reliance on fundamental business practices?

11.2 THE BUSINESS FAD GAME

The Business Fad Game is similar to many other games. To appreciate a game, you have to have a thorough understanding of its foundations. Take baseball, for instance. The dynamics of its play rests on three essential propositions: (1) good pitching dominates good hitting; (2) a sane player fears an inside fastball; and (3) there is something very Zen about the 90 feet between bases. The good-pitching fundamental has been proven over the long history of the game. Whenever pitching talent wanes, the game degenerates into a slugfest—an ugly form of the game attractive to some fans but distasteful to baseball purists. Baseball fans always pray for a new crop of good pitchers to keep the game in balance. The next fundamental, fear of the inside fastball, reinforces the good-pitching fundamental. Our natural tendency to pull away from an object heading toward us at 90 miles per hour, coupled with a pitcher's skill to place it within inches of our elbow or head, help the two fundamentals work together. The last fundamental, the Zen of 90 feet between bases, ensures that the few times a batter who does make contact with the ball will, most likely, make it to first base a half-step after the ball does. It so happens that 90 feet also seems to be the perfect distance to the next base, keeping a runner from advancing between the time the pitcher is committed to a pitch and the catcher can deliver it to second. All three fundamentals are necessary for the game to be simultaneously dynamic during play and stable over the years.

Like baseball, the Business Fad Game has its own three fundamentals: the Possibility of Success, the Security of Fashion, and the Law of Averages. The Possibility of Success lures management into signing up for the latest guru's program because such cures, when successful, supposedly make the company a lot of money. It is easy to convince yourself that a potential 80% failure rate is tolerable if the upside is sufficiently attractive.

Of course, the downside has to be protected. Given that most business fad programs will surely fail, the game relies on the second fundamental to minimize the impact of a loss: Security of Fashion. If "everyone" is signing up, it is easy to rationalize that there must be something to this new program. If you fail, you will feel safe because you will be in the company of many of your industry peers wearing the same fashion.

The second fundamental shows up in polite business conversations that go something like this:

> "So, how is your LMNOP Program going? What? You're not doing LMNOP? Everyone is doing LMNOP! I read just last week that Amalgamated Armotronics improved their quality 200% in six months and won the LMNOP Award. Their new president got a huge bonus. Didn't you see him featured in this month's issue of *Dysfunctional Trends*?"

Any red-blooded manager would not be caught dead without an LMNOP program. No matter how hideous the latest fashion looks, the manager wants it. After all, everyone's wearing it.

The third fundamental, the Law of Averages, ensures that the downside consequences of the game go unnoticed or can be explained away, on average, the same way a statistical average hides individual performance. A product of fashion has to be relatively harmless to be sustainable. Sure, some men with robust waists should not wear the latest European-style bathing suits, and some women past their prime perhaps should forego the latest skirt style. While certain individuals may look foolish, the harm such fashions cause is limited, superficial, and lost when we calculate the overall impact on humanity.

Likewise, the Business Fad Game is sustainable because we think that any harm that may come from it would be minimal, sort of like being caught wearing casual dress to a cocktail party. The Law of Averages says a game like this must *appear* harmless on average. On average, most managements feel obligated to try something to improve performance and profitability. On average, if the staff were not involved in one improvement program, it would be involved in some other improvement program. While fads may lower efficiency in the overall economy, they are a part of the way we do business. On average, you do not miss wealth you never knew you had. However, on an individual company basis—the place where each one of us lives—following a fad can be devastating. Unlike a friendly, entertaining parlor game, the business fad game has serious individual consequences. That is where the symbolism between games and business fads breaks down. People do not lose their jobs by losing at Parcheesi.

11.3 BREAKING OUT

Unless you, as a manager, find a way to break the business fad's life cycle, your future will be dotted with endless attempts at installing the latest program that bubbles up from the business fad industry. Laying aside the question of which particular programs might be right for you, you first need to assure yourself that there is a way to separate the few successes from the many failures. As we discussed early in this book, no fundamental change will occur if the threat to your company is vague and lifeless, if insiders believe that your company is inherently better than its competitors, or if you remain convinced that you and the people around you are doing the best job possible.

Manageering's third fundamental skill, understanding change dynamics, does not say that it will be hard to change if you lack a clear, overwhelming threat—it says that change will be *impossible*. You have to face near catastrophe and a clear, identifiable enemy before significant change occurs. Then, after you survive the war, you will need to have enough strength left to rebuild before the next onslaught. For example, take IBM's well-publicized troubles of the early 1990s. Many outside the company recognized the signs of crisis years before, as did a few insiders. The board eventually had to bring in an outsider. Only someone like Louis Gerstner could paint a stark picture of reality for the majority of insiders who had dozed off listening to stories of ancient success. Although his work at the time of this writing was far from finished, he admitted that he was lucky to have a real crisis to get people's attention and that IBM still had core strengths on which to rebuild.

The IBM story contrasts with the tale of Westinghouse. Concurrent and coincidentally with IBM's troubles, Westinghouse faced convincing evidence that something was very wrong with its basic business. Outsider Mike Jordan was brought in, but substantial changes were still a long way off after two years of effort. Jordan's plan had two fatal flaws: (1) insufficient attention to communicating the unequivocal need for change and (2) Jordan's reported conviction that the Westinghouse "culture" was the culprit and could be changed by seeding the headquarters staff with marketing and financial types. Their inability to communicate the depth and breadth of the crisis meant that the entrenched troops remained unconvinced that they had to change their way of doing business. The second flaw meant that Jordan defined Westinghouse's culture as the change objective. That objective implied that Westinghouse management was going to turn the company around by emphasizing a presumed weakness—the lack of entrepreneurial, market-driven focus—while ignoring a core strength—its technical bureaucracy. Not coincidentally, a corporate culture makeover was one of the change programs fashionable during the Westinghouse turnaround attempt. Unless it has a serendipitous run of good luck, Westinghouse will become just another story of business fads gone mad. If that luck comes about, however, corporate culture change programs will become all the rage.

Although always real, not every threat is as in-your-face as IBM's or Westinghouse's. The difference between success and failure of many improvement initiatives is not the change initiative itself, but the creative leaders implementing it. Those persons see the threat more clearly than others, energetically describe the faceless enemy to others, and engage battle before their peers would. Whether you find one of the popular, off-the-shelf business fads attractive or see all of them as last Sunday's leftovers, you yourself will have to make your change program work by using your intelligence, by applying the force of your personality, and by persisting via hard, almost never ending work following the framework of a set of fundamental skills like manageering.

11.4 WHAT IS IN A NAME?

What makes business fads both hard to kill and so perversely abundant? For some clues, let us look at how our business era is popularly characterized. You are constantly told that this is the "computer age" or the "information revolution." What is in a name? More than you think. Names have implied, unspoken meanings. Remember learning in the seventh grade about the "age of the steam engine" and its variant, the "industrial revolution"? The first variant (age of the steam engine) characterized the machine whose principal impact on society was embodied in the second variant (the industrial revolution). It is a visual mnemonic. See the steam engine, see industry.

Like the term "age of the steam engine," the term "computer age" refers to the predominate machine of our time. But the symmetry does not hold with its variant. Unlike the word *industrial*, as used in "industrial revolution," the word *information* refers not to the machine's impact on society but to the substance *in* the machine. It is like calling the "age of the steam engine" the "steam revolution," but for obvious reasons we do not. It is redundant. Then why call our era the "information revolution"? We should not. It is an impotent descriptor implying that content and impact are the same thing, which of course they are not.

11.5 THE FAD FACTORY

So what is the real impact of the computer's content on our society? While it obviously has lowered costs in many industries, in the world of business management it plays both a founding and a supporting role in the management fad industry. Fads are unsustainable unless they can quickly generate a critical mass of followers. To reach that critical mass, business fads need really fast and ubiquitous communications systems, which themselves depend on cheap computer technology. Furthermore, business fads need computers to provide efficient "information" cloning—the core manufacturing activity of the business fad factory.

Therein lies the danger of marveling at the machine rather than looking at the real work it performs. To appreciate the age of the steam engine, you need neither a grasp of the machine nor of thermodynamics. You need to know how the machine itself was used. What is important to understanding our age is not a comprehension of the computing machine nor the information in it, but an awareness of how its output is used. Future generations of managers will not marvel at the computer any more than you marvel at the steam engine. Most likely, when they look back at us, they will even forget to put the computer in the picture, just as you forget to put the steam engine in your mind's eye when you think about the great textile, steel, oil, shipbuilding, and railroad industries of the past. Your great-grandchildren will not blame the computer for our management failures any more than we blame the steam engine for the terrible working conditions of the late 19th century. History assigns credit and blame squarely on the shoulders of the *people* who applied innovation, not the innovation itself.

Your life and times are defined by the actual, real work you do, not by the machines you use nor the content of the material you work. It is the real value you leave behind that counts. Following a fad for fashion's sake leaves a sterile legacy.

11.6 IN THE BEGINNING

When did the era of the business fad begin, and how do you know you are in it? We need to go back to a previous era to understand how we got where we are today. You have read or heard about the golden age of the post–World War II business boom. Instead of reliving the myths of that era, let us review a few facts. New businesses were forming at an incredible rate because established businesses could not create enough jobs for all the returning soldiers and sailors. Because their war-time experiences taught them how to overcome almost any obstacle, many of those soldiers and sailors did what they had to do to make a living: They started their own businesses. They augmented their Army and Navy leadership skills with on-the-job trial and error. Many failed, but those who survived were quick, smart learners. They did not learn their business lessons from following *Wall Street Journal* articles about the trials and tribulations of Ford, Mobil, or General Electric. That world was irrelevant. Survival was the name of the game. At the time, business tended to be regional. If you had a business problem, you sought help from those around you. Local business leaders, your family, your customers, your suppliers, and (surprise!) your competitors were your best sources of information. A good idea was a good idea no matter where it came from.

Many of those post–World War II business leaders began to retire in the 1980s. Typically, they sold their companies to others who had "strategic acquisition plans." The new owners brought in bright, professional managers, who in many cases were trained or influenced by business schools. People like yourself. But American business schools by and large teach business like they teach engineering and science: by carving it up into semiautonomous disciplines like finance, marketing, and accounting. When they teach general management skills, they rely on the case method, a technique that leads to a management-by-case mind-set. When those ideas escape campus, they become management fads, infecting a lot of otherwise normal and even gifted people.

But we should not get carried away. Business schools are not the problem behind poor corporate performance any more than American engineering schools are the cause of poorly designed highways. However, business schools do have a tendency to reinforce a certain way of thinking. And that thinking, when turned loose on the world, looks for ways to validate itself.

Joining business schools on the list of business fad enablers is the general business press, an *eminence grice* that gives the fad industry a nearly everlasting life. Journalists face a relentless, all-absorbing, grinding problem of how to fill the white space or dead air between advertisements. Thoughtful consideration of tactical business problems is hard to communicate. Simple, monotonous patterns proffering the latest fad answers are easier to write, easier to edit, easier to repeat, and easier to swallow. The journalists' challenge is to sell papers, magazines, books, and television programs in an increasingly saturated, distracted market. Business journalists build a world of fiction similar to made-for-television, "reality-based" programming. They should be required to put a disclaimer with each piece, as Hollywood does on many of its products: "Any resemblance between the characters and scenes depicted in this work and any real-life situation is purely coincidental. This story is a total fabrication."

When pundits tag our life and times with terms like the "computer age" or the "information revolution," they are guilty of infecting us with provincial, inbred ideas and meaningless meanings. When we discussed manageering's fifth fundamental—decision theory—we addressed how framing influences understanding and limits choice. The feeble frames offered by experts give you no boost but instead stand in the way of breaking loose of the business fad industry. The next time you hear someone refer to your age and time in those terms, reframe it in terms of the tyranny of the business fad.

11.7 FUNDAMENTAL FOUNDATIONS

The business press is not the only vehicle used by the American business-fad industry, it is just the most obvious and most persistent one. Finding your way out of the maze of business fads and fakes takes more than seeing through such self-promotion. After all, the maze mutates into a new form every two years or so, adding a new corridor, mirror, or passage leading back to itself. Finding your way out takes fundamental problem-solving skills. You have to be able to look at the once-familiar suspiciously or at least in a different light. You have to be able to distinguish between two or more seemingly identical choices. You have to be able to back out of dead-ends without running into yourself going in. A map can be helpful, and you might suspect that this book is one. It is not a map—it provides a lesson in map reading. Maps are easy to come by, but map-reading skills are in short supply. Even skilled maze makers themselves become trapped in their own mazes, unable to get out even though they think they know everything about their creation.

What you need more than a map is cross-country trekking skills. You need to be able to find old, faithful, but seldom-used paths. You need to know which places to avoid even though they are not marked as dangerous. You need to spot false turns, dead-ends, and drop-offs before you reach them. An accomplished trekker knows that few maps show that kind of detail. A map is only a start; a successful journey depends on decisions you make after leaving civilization behind. We hope that this book has helped you become that experienced, knowledgeable, and mature trekker.

Back in this world, general managers, presidents, CEOs, and others managers avoid the traps of the fad maze by relying on fundamental management skills, whether they realize it or not. Those fundamentals do not include quality program skills, although the quality of your product and service hardly can be ignored. They do not even include team-building skills, although building teams is a necessary activity. Reengineering or right-sizing is not on the list, although improved operational efficiency is the tangible result of applying fundamental management skills. Instead, fundamental skills allow you to see through the maze to extract the real, underlying, and basic business ideas in each topical management salve. It is time for us to reject the fad mongers in favor of more critical skills. The legacy of your generation of business leaders depends on it.

11.8 HALL OF SHAME

Now that we have examined the dynamics of the fad industry in America, we will gather together some of the more popular management fads of the 1990s and see if we can make any sense of them.

Our treatment may seem a little harsh, especially if we pick on one of your favorites. We have no malice toward any one fad. We believe that all of them are universally flawed, not because of their basic premises but because their practitioners tend to elevate one fundamental over all others. We want to show how the complete set of fundamental management skills presented in this book, along with a little common sense, can reveal that the emperor has no clothes.

But before we go about trashing some the currently popular management fads, we have to answer two common questions. First, given that most American products built in the 1990s were better than those built in the 1970s, should not the management initiatives associated with this era share some of the credit for that success? Second, how can so many people be so wrong?

While we agree that products built today are, in general, better than those built 20 years ago, it is also true that products built in the 1890s were substantially better than those built in the 1870s. That era was able to improve on product design, manufacturing methods, and organizational structure without the benefit of management fads. Therefore, how much of the improvement we saw during our lifetimes should we attribute to the various late 20th century management fads? More to the point, is not it just as likely that we would have had *more* improvement had our taste for management fads been subdued? We would like to find a recent study of the impact of management fads on the American economy that shows that fads have had a negative impact on our economy. But we are resigned to the fact that if such a study existed, it would be flawed because no one can predict the path not taken. In the end, you have to judge for yourself whether or not fads, per se, add to our GDP. It is an article of faith, based on personal experience, extensive observation, and ruthless comparative analysis, that fads have little redeeming value.

The second question implies that mass delusion is a rare event and that a measure of an idea is its popularly. World history is full of instances in which mass delusion was commonplace. In the late 1600s, the European aristocracy was caught up in the Tulip Craze, a speculative bubble that saw the price of tulip bulbs exceed thousands of dollars (in today's dollars) each. Among those caught by the market's eventual collapse was Sir Isaac Newton. More recently, hundreds of millions of dollars were invested in the early 1980s in the domestic oil and gas business based on the widely accepted forecast of $100 per barrel of oil and $10 per 1,000 cubic feet of gas. Since then, the market price of oil has rarely gone above $20 per barrel and gas above $2 per 1,000 cubic feet. On a sadder and recurring note, popular belief is the root of genocide. Hitler's *Mein Kampf* was a worldwide best seller and formed the basis of one of the more horrific examples of the error and terror of the masses.

On the other hand, as we have pointed out elsewhere in this book, some people somewhere at some time get some benefit from pursuing the latest fad. But a review of the drivers of change indicates that almost any management initiative would have worked given the unrelated but fortuitous circumstances of those who succeed. It is not the fad; it is the intent and the talent of the people who want to change. It is our experience that most management fads are born *after* the first-mover moved. First-movers rarely interpret what they have done as an innovative management initiative but rather as a response to pressing needs. Only after the business press picks up on a particularly good business success story does a management fad spring up around a set of practices identified with the latest corporate hero. Whether other people can benefit from those collected and associated practices depends more on good luck and dedicated management than on the power of the subsequent management fad.

11.9 REENGINEERING AND TECHNOLOGY

For our first analysis, we will look at reengineering. At the time this book was written, reengineering had just entered intensive care, and its slow death was at hand. Reengineering is a paradigm for the business fad movement, not only because its fast rise and short history were so typical of management fads but also because it championed a cherished myth about the power of technology.

Reengineering had two advantages over competing fads of the day: fortuitous timing and superficial understanding. As for timing, it was introduced during a coincidental spasm of widespread, widely reported layoffs (euphemistically called "right-sizing") that was thundering across the American business landscape. Reengineering promised that your company could do the same amount of work with fewer people, but only if it followed certain principles of management and systems design, dubbed *reengineering*. Superficial understanding came into play and contributed to its success as a business fad because managers implementing reengineering neglected to read the seminal book, *Reengineering the Corporation,* by James Champy and Michael Hammer. Central to the authors' thesis and indivisible from it, was the role of technology in successful reengineering programs, particularly computer technology. In the end, reengineering became a buzz word rationale for mindless downsizing. The point is not that downsizing was or was not needed; it is that superficial managers, who read only the book reviews, directed their staffs to "reengineer" the company by cutting staff some arbitrary amount. The confusion over what reengineering was and was not became so laughable by 1995 that many confused it with TQM—an older, more resilient fad that we examine later in this chapter.

Every management fad has a quick-fix theme. For reengineering, technology was the holy grail. In Chapter 4, we said that the impact of new technology on business transformation has remained about the same throughout recorded history. While the technology we fixate on today is the computer, such machines are by no means the only new technology presenting opportunities for profit. However, computers seem to have captured the imagination of pop culture journalists in a way not possible for plastics, metals, ceramics, and packaging materials, all examples of emerging technologies that may improve our standard of living by lowering costs.

The computer industry has done an extremely fine job of selling its technology as not only essential to survival for all industries, but practically as essential as air and water is to life. However, you do not have to dig too deeply to find disconfirming information concerning the value of computer technology in running businesses. Before record downsizing hit American business in the early to mid 1990s, economists were perplexed by consistently repeatable numbers showing that the productivity of American office workers was stagnant, despite the billions of dollars spent on automating overhead functions. There were bright spots here and there and good stories to tell, but no matter how it was measured, white-collar productivity did not seem to be getting the same boost from computers as the factory worker had. What were people doing with such large capital investments in the front office?

11.10 FADS WALK ON THE BACK OF TECHNICAL ILLITERACY

The central problem with employing computer technology in the workplace and, therefore, with reengineering in general is that managers approving the implementation of the technology are technically illiterate. They give too little thought about how technology changes the way people work. Without any concrete idea of which jobs will change, where the efficiencies will come, and exactly how people will do their jobs after the new computer system is up and running (as opposed to how they did it the day before), costs do not fall. Rather, many technological applications lead to additional features, which on examination add little value. "You will be able to get your reports on a daily basis instead of weekly," you are told. What good is that if you pile them up on your desk and continue to review them on a weekly basis anyway? "You will be able to model the market much more accurately," you are promised. Says who? We do not even understand how our models work today. Why should we pay more attention to one generated by a machine when it cannot even predict the past? "With e-mail, we will be able to improve communications," you are told. We are covered with meaningless memos and weighed down by voice mail already. As if I need another way to avoid talking to someone directly!

All these snide remarks cut to the quick. Applied technology requires a lot of preinstallation thinking about how the technology will improve profitability and change work routines. For example, suppose your product is metal-based, and you are exploring the possibility of replacing some of the metal with plastic. Would you call in a few plastic vendors, get their ideas, and then order a couple of tons on the bet that your operations staff could figure out how to use it? Of course not. Then why are computers and computer systems bought that way? Our experience is that rigorous operational analysis and cost justifications for computer investments are seldom done because many general managers are illiterate when it comes to systems design and fundamental problem solving. Few seem eager to learn, and vendors capitalize on that attitude and ineptitude.

A good example of how inattentive use of technology can overwhelm technically illiterate managers is the problem of excessive e-mail. In the mid 1990s, general managers realized that e-mail provided new ways for diverse groups to communicate, especially those whose work was interdependent but separated by thousands of miles and tens of hours. Regular mail was too slow, overnight charges too expensive. E-mail provided the right mix of service and cost. Eventually, however, e-mail traffic grew so fast that getting through your daily e-mail before the end of the day became problematic. Messages started piling up, unanswered. E-mailers began habitually telephoning or faxing their intendents to tell them that an e-mail message was waiting for them. The conventional answer to this emerging problem was to use technology to rescue what was basically a nontechnological problem. Applying declassified intelligence processing techniques, software developers promised programs that would digest e-mail messages and, using artificial intelligence, produce summaries so you could decide which messages needed to be read in full and which you could, presumably, just pretend that you had read before you junked them.

The solution would be absurd if it were not so shocking that many managers actually thought that this was a good solution to the problem. The real problem was not with the receivers' inability to read their mail; it was with the senders' inability to write. When our business skills become so poor that we cannot depend on people to write cogent summaries of their own work, when we become such poor teachers that people cannot stop themselves from using the word "subsequent" seven times in two pages, when we become so lame that we lose sight that general management—not technology—is responsible for designing command-and-control systems to regulate communications traffic, it is time for us to go home.

The problem with uncritical application of technology is profoundly difficult to overcome because today's technology is cheap to acquire and widely available, compared to the technology of our grandfathers. The best hope for avoiding inappropriate application is by becoming sensitive to it, not by becoming antitechnology rubes. Being anti-technology is as lame-brained as embracing technology thoughtlessly. The questions you ought to ask each time any new technology is proposed is, "How will this investment in technology change the way we work? How will that change alone make us money?" Those questions will not keep you from making bad technology investments, but they will lead to other questions that lead the way to a better understanding of what you are about to undertake.

The central problem with unsuccessful technology usually comes from adding significant cost without appropriate gains. A typical situation that leads to such an outcome is when you find yourself supplanting craft with a technological fix. Sidney Lumet, in his 1995 book *Making Movies,* makes the point that the addition of Dolby and other exotic sound techniques has appreciably increased the cost of film making while becoming de rigueur for all films, whether or not the technology adds to the movie experience. Do we really need Dolby and SurroundSound for a 19th-century art-house picture? Obviously, someone is replacing the craft of sound creation and effect with hot technology without asking whether the product gains anything from the effort.

A tale that illustrates the difference between blindly accepting the surrounding hype and understanding the way a new technology will actually change the way we work comes from events closer to home. It is the familiar story of how availability of faster computers and cheaper data storage has destroyed the elegance of a common and accepted business solution that still had value but was discarded as immaterial to the new world. Contracts used to always be dated at the beginning of a month, and any partial months resulting from being off-cycle were handled through a simple, one-time prorated charge. Why? Because by doing so you simplified the receivables-aging calculation, a considerable task for even medium-sized companies.

With the advent of computers, naive business managers abandoned even-dated contracts since everyone knew that computers can count. If the collective business community had bothered to ask how the work was going to change, they would have realized that the mechanical work—that of adding numbers and sorting—was the only thing that would change. Intellectual work would remain unchanged. The basic problem of defining how long a standard month and a standard year is for commercial relationships that begin within any month described by a nonmonotonic calendar is still with us, no matter what calculation tool is used. What was forgotten when business embraced the computer was that the core aging problem does not get solved just because a computer intervenes in the work. The intellectual work actually became more complex because what was thought to be an ancient and unneeded solution to an obsolete problem actually will always be with us as long as we use a Gregorian calendar. The real work of aging receivables has not changed.

If you do not ask these types of questions and struggle with their implications, technology becomes an expensive fad. Stories abound how premier companies like Federal Express and American Airlines used information technology in the 1980s and 1990s to their advantage. What is forgotten in the retelling of the stories is that their new technology was part of a coherent operational strategy that centered on that technology. Too many companies assign the care and feeding of new "technologies" to an existing function rather than looking at it as an integrative solution that remains the responsibility of general management.

11.11 THE KING OF CONFUSION: QUALITY IMPROVEMENT PROGRAMS

What would happen if the chairman of the board of Amalgamated Armtronics, a publicly traded company with $20 billion of yearly sales, sent out a press release saying that they were embarking on an innovative program that included such novel ideas as watching costs, protecting market share, and generally listening to customers? Your retort might be, "What have you been doing up to now? Ignoring costs, sales, and your customers?" Of course no one announces an improvement program that emphasizes normal business behavior. How is it then that companies routinely announce with pride that they are embarking on a "quality improvement program"? Was quality a feature ignored before and just now noticed? More likely, they are just aping the latest fad.

The first indication that quality improvement programs are destined for trouble is that their objective, quality, is an undefinable term. Any effort aimed at improving quality runs into endless debates concerning what quality is and what it is not. The problem is so severe that leading practitioners always start their canned quality program by defining the term, only to be attacked by other equally expert practitioners who loudly proclaim that their definition is better. This intellectual arm wrestling sets the stage for continuous turmoil in the prepackaged quality improvement program industry.

This problem will never be solved because all programs organized around quality themes harbor an inescapable paradox. Quality program advocates generally express the problem as one of definition, implying that inconvenient semantics lie at its roots. It is not a problem of semantics. It is much deeper and more profound than that. To understand how devastating the problem is, you need to know something about a fascinating mathematical discovery called Gödel's theorem. (You do not need to know anything about math to appreciate its implications.)

Some biographers and historians say that the great philosopher and mathematician Bertrand Russell became clinically insane or at least a broken man because his life's work was negated by Gödel's. If such a premier mind as Bertrand Russell can be crushed by this paradox, think what happens when mere mortals in the quality improvement industry find themselves floating in the same intellectual sea. Russell spent nearly 20 years of his life proving that mathematics was the one fundamentally pure science. He attempted to do that by showing that its foundation—logic—was consistent, coherent, and self-sustaining. The hallmark of such a system is that it is free from paradox. Paradox in mathematics or any system appears when a statement is unprovable from within the system, but is nonetheless true. We face paradox in many human endeavors, and some say the power of the human mind is that it can cope with these unprovable truths. How we do this is still unknown and may be unknowable.

Mathematics was supposed to be free of paradoxes. Mathematicians assumed that every and all mathematical hypotheses can be proved either true or false, given enough time, insight, and creativity. Russell wanted to prove that assumption. He wanted to prove that mathematics was unique in that it could support itself from within, exclusive of any help from an outside observer. That is not a trivial point. All science and rational human progress depended, as far as Russell was concerned, on mathematics having such a basis. It drove to the heart of the question whether mathematics existed as a universal truth or was a construct of the human mind.

In relation to our own daily lives, Russell's effort seems like entertainment for eggheads. We take for granted that math is math. Anywhere in the universe, $2 + 2 = 4$. Gödel's theorem (called the incompleteness theorem for reasons that will become obvious) removed mathematics from the realm of truth and put it in the realm of faith. More than that, it removed any hope of escaping paradox in any system, whether it be math, physics, or quality. In essence, Gödel said that any system built on a set of axioms (i.e., statements taken as true on their face) will have an indefinite number of statements that cannot be proved true or false but are nonetheless true. Furthermore, this troubling feature remains no matter how many additional axioms you add in your effort to nail down "truth." The problem rests with nature's apparently self-referential nature.

The current and forever continuing problem with quality improvement programs is as close as most of us will get to see the effects of the incompleteness theorem. It is more obvious in the quality field than in mathematics because the self-referential nature of quality is not hidden beneath layers of thought but lies on the surface. Most technologists use mathematics without becoming overwhelmed by this esoteric problem. Not so with practitioners of quality. The very foundation of their work rests on the obvious self-referential concept of quality. We say something is a quality piece of work, or that it reflects quality, never worrying that we are referring to a concept that is unprovable but nonetheless true. We know that quality exists. If we were to remove the concept from our world, the world would be a very different place. But just because something like quality exists does not mean we are able to define it.

Before now, you may have been unaware of this small but significant tear in the fabric of the universe. Maybe you thought the problem was yours alone when you became vertiginous every time you did quality thinking about quality. But if you lose your balance when thinking about quality, think about the kinds of problems you will have when a whole company takes up a program built on a self-referential concept. It invites chaos. Because of the limits of self-referentialism, we have to accept on faith that quality does exist. We cannot define it sufficiently to say that we have a "quality improvement program" in place. That is somewhat equivalent to saying we have a "business program" in place.

This fatal flaw in quality systems development does not stop people from selling quality improvement programs, and 30 to 40 percent of targeted companies feel that they have accomplished something. In the practical, self-sustaining world of quality consultants, some experts limit their definition of quality to "conformance to requirements." That neat trick serves to convert a quality improvement program to a conformance-to-specification program. What is left out of this neat conversion is the troublesome fact that someone outside the "quality system," but presumably having access to it, has to define what the requirements are. We assume that they depend on their own personal sense of what quality is. Some other practitioners attempt to solve the problem by assigning the responsibility of deciding what quality is to the customer. Aside from the fact that someone inside the company still has to interpret a customer's self-referential statements about quality, we will show you next that in the case of quality the customer is not always right.

Feeling a little off balance now? That is the same feeling, but less intense, that overcomes anyone involved for any length of time in a quality program. It is the feeling that you are making progress only to find yourself at the same place you were earlier. Lots of scenery has passed by, but all you have done is burned up some gas, depreciated your car a little more, and engaged in sometimes boring, sometimes interesting conversation. You still have not made any money, the first fundamental of manageering. That is not to say that some people, somewhere, sometime actually get value from a quality improvement program. They certainly are not shy about sharing their success stories. But underneath the story is the truth that certain elements of their quality program were more successful than others. The successful elements are always the ones that fortuitously had a force-field profile (see Chapter 5) conducive to change at the time the program came along. The successful program elements fit well with existing and recognized operating problems. All the company needed was a methodology to convert the existing pent-up demand for change to action. The fact that the methodology was delivered under the cover of a fad was coincidental to what really happened at ground zero.

Ultimately, quality is not a thing but a condition that describes the net result of a total effort. You decide what quality means. A canned program to improve quality turns out to be a grab bag of tools, techniques, and slogans, all having their appropriate place and time. Many times, the tool is inappropriate, the time is wrong, and quality is not even the problem. It is the result. For instance, think of the times you go into a store and are disappointed with the result. You do not say to yourself, "Gee, that store needs a quality improvement program." You think in specifics, visualizing the need for enhanced training, attentive management, newer equipment, designed systems, faster processing, better whatever. You might say you were disappointed in the "quality," but it would not take long for you to explain your self-referential statement to your satisfaction.

11.12 THE CUSTOMER IS SELDOM RIGHT

Before we proceed with examining a few specific fad quality improvement programs, like TQM and ISO 9000, we need to return to our claim that the customer is not always right. The myth that the customer is always right has done more damage than many of the bigger-name fads.

Years ago, we were fortunate to work with one of the media-proclaimed innovators of just-in-time (JIT) inventory management. He was a self-made, American-grown, original business thinker. He became wealthy by challenging conventional thought. What was obvious to most people was perplexing to him; what was obvious to him was downright inaccessible to others. One of the presumptions that perplexed him was that you should run your business as if the customer was always right. He figured since he was in business to provide an industrial product that his customers knew less about than he did, would not it be natural for his customers to assume that they did not know what they wanted beyond some very vague concepts? If they did not know the product well enough to build it, how could they always be right? Is not it more likely that customers are more often wrong than right?

The problem of ignorant customers is familiar to almost everyone in business. It is just impolite to admit it. We find more sales and customer support personnel who disparage their bread-and-butter customers than praise them. The conventional fad fix to that poor attitude is to "educate" your sales and support force so that they better "understand" customers and are more responsive to their needs. However, if it is painfully obvious that disdain for customers is the rule, not the exception, how successful can a program be when it focuses on the symptom instead of the cause? Quick fixes redefine this conflict as a simple problem of attitude. Fad programs have to do that in order to construct easily digestible, readily sellable solutions. It is simpler to say that your sales and support people need improvement than to admit your basic business plan needs rethinking.

Therefore, management fads organized around the customer-is-always-right myth are popular and seemingly successful. For example, in the mid-1990s, the Ritz Carlton Hotel Company won the Malcolm Baldrige National Quality Award in large part because of a proclaimed policy of customer righteousness. Ritz figured that, on average, the customers know better what they want than Ritz personnel and that, by enforcing the core policy of doing everything the customers asked, on average, they would increase customer satisfaction. Underlying that customer service policy was an assumption that increased customer satisfaction would lead to increased profits. The additional costs to do whatever the customers want presumably would increase their market share of the very upscale guest market.

Overlooked by casual admirers of the Ritz approach is that for the Ritz many customer requests are billable activities. If customers ask that their cars be washed and waxed overnight or for a special cuisine, it will be done, but they will be charged appropriately. The typical Ritz customer expects to pay for superior service. On the other hand, if a customer does not like some part of the service that is considered as included in the price of the room, then, of course, making it right is part of the agreed base price. On the other hand, if a customer has all sorts of unusual and expensive requests and refuses to pay for them, Ritz classifies that person as a "noncustomer." This philosophy—getting customers what they want for a price—is not peculiar to the Ritz. They just do it on a bigger, pricier scale and perhaps are better organized to handle unusual customer requests systematically with a highly trained service staff. The Ritz's actual policy on the ground is not that the customer is always right, but that Ritz employees will do everything they can to provide the service they promised to their profile customer. In the rush to emulate the Ritz, the simplifying fad transformation drops half of the equation: the need to identify exactly how a policy will make money. Instead, platitudes and obvious-by-inspection attitudes replace well-thought out, quantitative justifications.

This is not a subtle point. The assumption that the customer is always right presupposes you have already defined who your customer is. When this philosophy is put in practice, the second part of the equation is dropped because it is messy. In the real world, you cannot separate your customer profile from customer satisfaction and price points. For example, during a recent company function at a fairly expensive Texas steak house, dinner guests were relaxing after the meal. The discussion at one end of the table turned to the meal itself. All were delighted, except one person. She was disappointed, even though she got the exact meal and service the rest of the party did.

Asked why she was disappointed, she mentioned she expected more numerous wait staff, better table appointments, and generally a different atmosphere. There was no way the restaurant could have entertained her disappointments. They design their meal presentation and service around a target customer. She was not their customer even though she thought she was. That is not to say that the restaurant or any business can ignore a dissatisfied customer. However, your customer service policy must be able to differentiate between a complaint from a nonprofile customer and one from a profile customer. We will leave it to you to think about different ways you could construct a system that responds to that difference without aggravating noncustomers.

This kind of thinking is blasphemous to many of today's management fads that encourage followers to listen to the customer and instill the presumption that the customer is always right. They teach that the customers have all the answers and that they know better than you about all things related to your product and service. However, to say that you must listen to your customer is not equivalent to saying that the customer is right, even though the two statements are used interchangeably. In reality, as general manager of your company or division, you have to have your own ideas about how you are going to approach a market and how you are going to pick your customers. (Yes, contrary to conventional wisdom, customers do not pick vendors, vendors pick customers.)

11.13 TOTAL QUALITY MANAGEMENT (TQM)

Among the countless quality-improvement programs making the rounds in the 1980s and 1990s, TQM was by far the most durable. Starting out from humble beginnings in the inspection department, its first incarnation was as total quality control (TQC), which gained some level of widespread acceptance from Dr. Armand V. Feigenbaum's 1961 classic book of the same name. But quality practitioners had bigger plans for themselves. TQC seldom reached into the general management ranks, and without that backing it would never become the self-promotion vehicle that quality control professionals needed. As the quality function became more professionally organized, the old moniker of "control" did not seem to fit its promise or direction of emerging fad management programs.

The origination of the term total quality management is debated by people who argue those types of things, but it nonetheless satisfied two constituencies. Quality professionals continued to be disappointed at the limits of their influence and recognized that to improve the reject rates they were experiencing in their inspection function they would have to expand their reach and authority into other parts of the company. Regardless of the professional politics involved, they were right. The general consensus expressed in literature of the day pointed out that control was an out-of-favor term and really what was needed was involvement of general management in the quality function. But a funny thing happened. The trend has not been for more management involvement in the quality function, but more quality function involvement in general management. The vehicle has been TQM.

All the thunder and noise about TQM drowns out the genuine moaning of real people frustrated by their management's failure to attend to its responsibilities. As we have emphasized throughout this book, the care and feeding of the operating system is the sole business of general management. When that responsibility is parceled out to functions unable to influence behavior across the board, middle managers become prematurely gray. It does not take long for these trapped animals to realize that a TQM program of whatever flavor is the preferred (although temporary) escape route, compared to fighting their way out by telling the boss he is not doing his job. Aside from systemic problems caused by undefinable terms and unmeasurable criteria, TQM programs continue to fail because general management is not convinced that real change is necessary or, if it is, that the change starts and ends with them.

We have been referring to quality fads such as TQC and TQM without defining them. That is not an oversight, but we figure that if quality practitioners do not have a unified definition of TQM, we do not need one ourselves. A specialized publishing industry surrounds TQM, and depending on the current fad, TQM can take on various forms, permutations, and specialties. Like the word "quality," TQM can mean just about whatever you want. Besides TQC, other incarnations include the Deming method, statistical process control, the Juran Institute, the Crosby School, and the National Quality Awards.

11.13.1 Total Quality Control

The granddaddy of them all, TQC fell out of favor before it had a chance to be really big. An idea ahead of its time, it was replaced in the collective industrial mind by new, improved versions. In the 1990s, Dr. Feigenbaum's book was in its third edition and still in print. It is an excellent reference for the practice of quality control and is extremely readable and practical as a technical book aimed at engineers. It identifies the practice of quality control as it applies throughout the production cycle, from product development to delivery (thus, the inclusion of the word "total" in the title). Dr. Feigenbaum unified those distributed functions under the one function called quality but implied that the tools he recounted were for everyone, not just the "quality staff." Dr. Feigenbaum's one shortcoming, if he had any, seems to have been his lack of effective self-promotion.

11.13.2 The Deming Method

For many, TQM is embodied in the teachings of Dr. W. Edwards Deming, although Dr. Deming never, as far as we know, identified his methods with the TQM tag. His name appears on the Deming prize, a national quality award presented in Japan by a quasi-governmental bureaucracy charged with "guiding" industrial development in that country. Deming became a quality guru in America when the press decided that everything American was inferior and everything Japanese superior. The American media's early 1980s investigation of this crisis turned up Dr. Deming. They proclaimed him as standing at the headwaters of the impressive, authentic improvement of Japanese products. (It seems convenient that American journalists found the savior of Japanese industry to be an American.)

While in some ways Dr. Deming did play an important role in the Japanese industrial transformation, the situation was and continues to be more complicated than that. Nevertheless, the management fad industry found a new product in Dr. Deming's tools, techniques, and philosophy. His widely repeated association with Ford's quality improvement initiatives ensured a steady market for books written by others explaining Dr. Deming's somewhat eclectic collection of principles, called "Deming's Fourteen Points." An entire industry of consultants, working day and night, dissected and repackaged Dr. Deming's prognostications for quality improvement. He had two basic, but not new, messages. First, he emphasized that "quality" was the responsibility, if not the sole reason for its existence, of general management. He was continually disappointed that his audiences were predominately populated by middle managers since he knew that change would come only if the leaders of industry got religion, so to speak. Second, his approach to product improvement was process oriented. He seemingly single-handedly reintroduced statistical process control to American manufacturers, deemphasizing inspection as the cornerstone of nonconformance avoidance. He came down from the mountain to remind us that "you cannot inspect quality into a product." Once people started listening to him, he had many other things to say and wound up settling on his famous Fourteen Points.

11.13.3 Statistical Process Control

Dr. Deming's black bag was statistical process control (SPC). The predominate tool of his form of industrial medicine was the control chart, a statistical method for analyzing the output of a process to determine the inherent capability of the process. From such analysis, the production staff was supposed to identify special causes of production upsets, and management was supposed to identify the opportunity for improving the entire process through investment in better tools, technology, and training. For some people, SPC and TQM became synonymous since the rest of Dr. Deming's teachings were more difficult to transform into a quick fix. Fifteen years after the big bang of the American quality universe, several well-known and respectable consulting firms continue to be organized around such a limited TQM toolbox and do quite well.

11.13.4 The Juran Institute

Dr. Joseph M. Juran was a contemporary of Dr. Deming and participated in the early efforts to help Japanese industry recover from the destruction of World War II. The professional relationship between these two proclaimed pioneers has been variously described as friendly, cordial, cool, envious, and hostile. The truth is unavailable to us and (according to decision analysis) most likely something entirely different and immaterial. The reality is that while both men had interesting things to say about quality control and the role of management, Deming's ideas were the first to be converted into a widespread industrial-cultural phenomenon by the American management fad machine, and to the first mover go the spoils. While Dr. Juran has written many books on quality, his *Quality Control Handbook,* which first appeared in 1951, remains his most popular contribution to the practice and is considered by some to be the bible on the subject. Not as well written as Dr. Feigenbaum's *Total Quality Control,* it nonetheless seems to include every quality control subject in the world (literally), ensuring a wide market and making it a very thick and heavy book that every self-respecting quality manager should have.

11.13.5 The Crosby School

Philip Crosby was the right man in the right place at the right time. His book, Quality is Free, published in 1979, appeared at the same time the popular press discovered Dr. Deming. He wrote other books on various topics, including *The Art of Getting Your Own Sweet Way, Running Things,* and *The Eternally Successful Organization.* Mr. Crosby's talent was marketing his ideas through an adult education company called Crosby Associates, Inc. For a time, it seemed that every up-and-coming middle manager in America was sent to Crosby's Florida campus to be inculcated with Crosby's philosophy. Serious gurus with "Dr." in front of their names did not think much of his sloganeering and rah-rah approach to quality, but the market sure did. While he, too, had a hodge-podge of points to present, we mostly associate the reinvigorization of "cost of quality" and the slogan Zero Defects with his camp. Misnamed, the cost of quality really should have been called "the cost of nonconformance."

Crosby realized that the emerging quality profession would get nowhere fast unless it talked the language of general management. He set about teaching quality professionals and upper-management wannabes established methods for quantifying how much rework and nonconforming product cost business so they could return to their jobs and get the message across to the people who really made the decisions. Instead of trying to reform general management, as Deming was attempting to do, Crosby went about reeducating middle management. Because of his approach, Crosby was a big proponent of defining quality as conformance to specification, and for that he was metaphorically crucified by quality purists who spent their free time arguing about the "real" meaning of quality. Like many quick-fix quality initiatives, Crosby's methods had little new to say, but they used refreshing narrative and had an easy style that sold well.

11.13.6 National Quality Awards

Not to be outdone by Japanese bureaucrats, the U.S. federal bureaucracy, with the support of various large U.S. companies and quality-profession organizations, came up with their own measure of excellence, the Malcolm Baldrige National Quality Award. It is awarded every year to fewer than a handful of companies competing in categories that group them by size and nature of business. The Baldrige award has had its good years and its bad. Some were astonished the year the Cadillac division of General Motors won the award since other measures of conformance and reliability did not put Cadillac anywhere near the top. Another year, the award was presented to a company that went bankrupt shortly afterward. Nevertheless, the appearance of a national quality award provided a forum (and government money) for quality professionals to thrash out the exact meaning of TQM. Not only is this effort slowgoing, an overlapping international effort to define TQM was underway at the time this book went to press (1997). This group began their epic struggle by trying to define the difference between quality assurance and quality management and promptly bogged down. Presumably, they skipped the definition of quality since they convinced themselves that they had a handle on it. We doubt it.

11.14 INTERNATIONAL HOUSE OF PANCAKES IS A CONSPIRACY

In the late 1980s and throughout the 1990s, American business was nearly overwhelmed by something broadly called "ISO 9000." This term was a code for inviting an independent third-party assessor to examine objective evidence at your facility and then certify that the documented "quality assurance program" met the requirements of one of a series of international standards, collectively called ISO 9000. Uninformed journalists and instant consultants ran three-column headlines uncovering all sorts of European conspiracies (the governing body of ISO was in Geneva), ISO-based export requirements and restrictions, and vendor mandates for certification. The fact that certification was a commercial activity, wholly and independently separate from the quality standards themselves, did not prevent the management-fad industry from using fear to promote the newest wave of "you-got-to-do-this-or-you-won't-be-a-player" mania.

The real disaster with the ISO certification fad was that the value of the quality systems standards themselves was totally obliterated by commercial certification drumbeat. More to the point, the certification game actually had a negative value for American industry and unfortunately was entropic—once industry started down that path, it had no way to come back to the starting point. Quality systems certification eventually will resemble other various product certification programs familiar to industries that build quasi-regulated products. It will be susceptible to conflicts of interest, double-dealing, payoffs, fraud, and general bad-guy-type stuff. Customers will demand that their vendors have third-party certification of their quality systems, but the process will add nothing but cost. The sad outcome is not in question. We have plenty of industrial history and human nature guaranteeing it.

Buried somewhere under this avalanche of ISO 9000 commercialization lies the real value of operating system standards. Disregarding that quality may appear in their names, operating standards outline systematic approaches to organizing your business. Of course, anything you do to improve the operation of your company affects quality. In many cases, those standards can be traced to efforts organized by quality practitioners who attempted to define qualitative issues in terms of quantifiable, observable actions. They are helpful guides in their original form. As they become requirements subject to outside interpretation, they lose all their intellectual authority and become someone else's property. Such standards go by various names (e.g., quality assurance standards, quality management standards, good manufacturing practices) and are available from various sources (e.g., industrial associations, professional societies, large buying cooperatives).

Operating standards can embody good, vetted ideas about generic operating systems, but as quality professionals seek to expand their industrial influence during the various revision cycles, groupthink will rule. For example, national quality awards had their beginnings in generic operating standards and are a good source of operation system design ideas. However, during the mid-1990s, they began to include more politically correct but immaterial directives concerning employee empowerment and organizational design. That creeping tendency to take basic, generic models and modify them to include "accepted" models of corporate behavior means that a good grounding in the six fundamental principles of manageering presented in this book is necessary to avoid swallowing some of the more contrived elements.

11.15 CULTURE CLUB

Given the problems and considerable effort required to implement well-defined operational changes, imagine the kinds of problems associated with grandiose schemes to change a company's culture. Applying the fundamentals we have presented, how does a cultural change project stack up? Poorly, that is how. First, if you thought defining quality was difficult, try defining culture. Second, suppose we pretend that you can generate a reasonable definition of culture. How are you going to measure your progress toward transforming the culture you have into the culture you want? How will you know when you have arrived? How do you know for certain that the new culture is better than the old one?

Most of what passes for culture is really cultural artifacts. Just as it is difficult to draw inferences from archaeological artifacts, it is almost impossible to say something intelligent about a company's culture by looking at its artifacts. Such immensely difficult problems trip up people trained in the science of sociology and anthropology all the time. They have raging debates about culture and its affects on those residing in it. Any fad management initiative claiming to change company culture is a fraud. It cannot be done, should not be done, and must not be done. Culture, like quality, is what you have after the day is done. It is not something you can manipulate, manage, forecast, and change. The time you waste pretending that you are improving your culture could be better put to use improving your product, understanding your market, or enjoying a day at the beach. Speculating on how to change your culture is a nonstarter unless you are talking about changing the company that pays your salary.

11.16 MY BENCHMARK WILL NOT RETURN MY CALLS

The problem with benchmarking is twofold. The methodology is internally inconsistent, and if that is not enough to bury it, it assumes that the past is prologue. The past is prologue if we are talking about the sweep of history; it is nearly irrelevant on the level of company operations.

Initially, benchmarking appealed to companies at the bottom of the heap because it supposedly pointed a path to the top. But what if you are at the top? Benchmarking does not help you. In fact, the success of benchmarking depends on getting good information about competitors or, in the absence of competitors, trying to find "similar" industries to benchmark against. Since companies at the top of their market have little competitive incentive to share their true competitive advantages, benchmarking databases could become dangerously skewed. People selling benchmarking information try to overcome that problem by appealing to the vanity of market leaders. They convinced some these leaders that benchmarking information could be helpful in determining how far ahead the leaders were. Nothing like patting yourself on the back.

But wait. If benchmarking is not such a bonanza for the guys on top, why should it be good for guys on the bottom? The essence of benchmarking is that you should copy the practices of the best. If the best do not benchmark, should you? As soon as you show that benchmarking is not so good for one segment, the whole intellectual premise supporting it collapses.

Suppose that this fatal flaw did not exist. Let us try another. Why should you follow a fad that focuses on the past if the consensus is that you should focus on the future? (Okay, we are relying on consensus thinking, which we have pretty much said is unreliable, but then we are only having fun here. We have already proven that benchmarking is a worthless science.) For example, in 1996 Ford introduced an entirely reengineered Taurus based on the direction that Ford thought Toyota was taking their Camry. Surprise! While Ford was chasing Toyota (in essence benchmarking against the Camry), Toyota went the other way. The 1997 Camry was scheduled to move toward fewer options and lower cost, while the Taurus was becoming more feature laden and more expensive. Time will tell which strategy paid off, but if Ford succeeds it will not be for its forward-thinking planning, nor will it prove benchmarking's worth. After all, perfect benchmarking, in which time lags do not play a factor, would have resulted in Ford moving down the features and price curve, not up. The impotence of benchmarking is obvious to any rational thinker, but as a fad it replaces thinking with fashion.

11.17 BUSINESS FADS DO HAVE VALUE

While these discussions may seem a bit sharp, they have to be. Our point is not that fads have no value; it is that their value can be buried so deep during conversion to fad status that it becomes all but obscure. Much of the problem is rooted in the absence of critical analysis by managers who lack a grounding in any set of fundamental management skills.

Our list of management fads is heavily weighted with quality-based fads because at the time of the first edition of this book quality was king of the fads. It was not always that way; every generation born during the era of the fad has its favorite path to quick riches and glory. When this book is updated a few years from now, only the names will change; the impotence of fad management movements will remain.

11.18 FADS AND THE NEED FOR CLOSURE

For our last bit of irreverence, we would like to discuss endings. Many of today's popular management initiatives conclude by saying that they never end. We are supposed to believe that their wisdom is so profound and their tools so powerful that your life will be influenced forever. That may be true. A management program initiated at the right time can have a disproportional effect on some people. However, the reality for most of us is that management programs do end, usually by being replaced by yet another program, and that is not necessarily bad. The assumption that a change program is never ending because change is never ending does a gross injustice to the reality of change dynamics. The nature of any program is that it must reach closure for the participants to feel a sense of accomplishment. Otherwise, no one can build on past experience. Change requires closure.

While some fad management programs do not provide closure because they do not understand the basic elements of change, others avoid discussing closure because they cannot imagine what would replace them, and therefore they cannot come to grips with the final chapter. We can. We want you to think about the six fundamental skills. Can you list them aloud without referring to Chapter 1? If you cannot, you are fairly average, and therein lies the point. The difference between successful managers and mediocre ones is that the successful ones recognize their limitations and fashion their own programs to do something about those limitations. Sure, you might hire a consultant to help build a proprietary program, but the program is yours. It will have a starting point, a middle, and a defined end. The fundamental skills for identifying the valuable elements of your program and for setting profitable objectives are here in this book anytime you want to revisit them.

As you try your own improvement project on yourself, you probably will emphasize one fundamental over another. After all, you cannot work on every facet of your skill set every single day. You have to take a break once in a while; you have to get closure on your latest efforts; and you have to revisit some of your earlier failures to get inspiration for your next move. We wish you the best of luck. We know how hard real change is.

11.19 MANAGERS ARE NOT THE ONLY FAD MONGERS

Before we leave the world of fads, we would like to leave you with a view larger than just business management by showing that the management profession is far from the only occupation susceptible to faddish behavior. During the mid 1990s, psychology suffered through another of its periodic molts. The public feasted on stories of psychiatric patients recounting satanic ritual abuse with the help of certified therapists using hypnotic techniques said to uncover repressed childhood memories. The number of cases uncovered could be explained only by assuming the existence of a widespread, organized satanic conspiracy in America. Despite the fact that no one could produce physical evidence showing that these cults ever existed, the practice of using elicited repressed memories for diagnosis became accepted treatment in clinics that were organized specifically for treatment of this new mental disorder. Repressed memory therapy, especially that which uncovered heretofore repressed episodes of long-term sexual abuse, became the profession's newest fad. Like fads of all persuasions, this one hitched a ride on the back of orthodox practice. Legitimate cases of memory repression were documented, and success rates were sufficient to allow its illegitimate application by questionable practitioners.

The common thread between medical treatment fads and business fads is how fringe practices become mainstream. The profession makes a small home for it in the name of tolerance and curiosity, only to find that with time it claims widespread acceptance. Recognizing that fads are not unique to business brings some comfort, but not enough when we consider the damage they do to the economy as a whole. Moreover, on a personal level, they can be devastating, and not just because people lose jobs when their company wastes precious time fiddling with fads instead of working on its problems. Even if a company survives and hangs on to its people, the psychic damage is considerable. A career peppered with meaningless work driven by a never-ending parade of business fads is a poor portrait for reflection during your twilight years.

We know that many people are overwhelmed by the faddish behavior surrounding them. It is hard, if not impossible, to run in the opposite direction of a mob. People have lost their lives resisting the rush of a panicked crowd. If you felt that way before reading this book, we hope we have given you the tools to dig out of this avalanche or at least enough air so you can hold out for the rescue team. If you find yourself in a management team shot full of fad fever, your time will come between the fading of the current fad and the blossoming of another. The cycle is inevitable, and the lull seems to come once every two to three years, so it will not be long. In the meantime, you have to shore up your own intellectual foundations so that when sufficient disconfirming information is available, you can use it to encourage real discussion about change, improvement, and profit.

Chapter 12

Closure

Every reader who gets this far made it here for singularly personal reasons. A few found professional inspiration in the book. They were looking for something to help them get over a hump and, through a fortuitous turn of events, found themselves holding this particular book. Others, maybe more managerially mature, feeling less a need for change and more a desire for information, found it a fun read. For them, it validated much of what they have been thinking about. Still others, perhaps having been forced to read it as an assignment, have been surprised that it actually had something insightful to say about how good business managers might actually work. But regardless of how or why you have gotten to this point, you share a common problem with all readers who get this far: You soon will forget much of what you have read.

We cannot change that, but what we can do is remind you that, whether we like it or not, repetition and application are the best teachers. That is why ideas like ours are reduced to paper: to allow us the luxury of coming back and revisiting ideas that, when read for the second time, become more meaningful than when we read them for the first time. The best use of an already read book is to reread it, but only when its time has come again. It is sort of like a good movie. After some time has passed, you want to see it again. You find yourself rewarded for the effort because you always find something new and meaningful.

Before you put this book down for the first "final" time, let us review the six fundamental skills of manageering with an eye not for repetition but for how you might actually encounter them again in real life.

12.1 PROFIT AWARENESS

As with most newly learned skills, the first fundamental you will forget will be the cardinal one. In golf, that is "keep your head down." In baseball, it is "keep your eye on the ball." In business, think "profit." Why is it so hard to remember the cardinal skill? For three reasons: First, when we are trying to learn a difficult and complex task, we try to do too many things at the same time. Second, profit awareness does not fit neatly into a manager's toolbox like the other fundamental skills seem to, so it is easy to leave it out. Last, most of us have few role models to help us develop profit awareness.

These problems imply solutions. First, we would suggest that you set aside a specific time period, say the next month, in which you make a conscious effort to recall the profit motive in *all* your business decisions. Do not even try to use the other five fundamental skills, just goose the gain on your profit awareness meter any time you face a managerial problem. Of course, after you recall profit, you naturally will move on and use the other fundamental skills during your problem-solving cycle. But by making the profit motive the first conscious consideration for every problem over a long period of time, eventually it will become automatic. Like batting practice in baseball, you have to set aside a block of time just for swinging and swinging and swinging at lofted balls to develop your "natural" tendencies.

12.2 TECHNICAL LITERACY

The challenge of gaining and maintaining technical literacy is more difficult than you might imagine, both for the technically competent and for the clueless. General managers coming from hard technical backgrounds suffer two special challenges. First, they are inclined to maintain their same high level of technical competence even though the expansion of their managerial responsibilities decreases the time available for such continuous technical enrichment. Second, they forget that the goal of technical literacy, as a fundamental management skill, is not technical competence. Technical literacy insinuates that managers shift their technological perspective from one of depth to one of scope. That is, instead of feeding their old technological roots so they grow deeper, managers have to spread their technological roots to tap into other people's technologies. Very few people can meet the demands of technical literacy and technical competence in the same lifetime.

Perversely, nontechnologists have an advantage in learning technical topics appropriate to general management because they carry little professional and educational baggage. They may begin with a few preconceived ideas about what the dominate technology for the company should be, but if they are mistaken they can easily displace their paradigm with the appropriate one. The disadvantage for nontechnologists is that they have a destructive tendency to question their technical aptitude and abilities.

The best way we have found for all managers to maintain a broad technical competence is to attend the same technical forums their technical people attend. Too often, general managers (as well as many staff managers) send their best and brightest to the latest offerings while telling themselves that they have no time to go themselves. Besides, they are afraid of looking stupid since they know they will not be able to understand everything that is going on. That kind of thinking misses the point entirely. Managers who go to technical conferences should expect to have a different agenda from that of attendees who come from the technical staff. First, they should expect to learn what the emerging issues are without trying to pick up on every nuance. Second, they should expect to increase their appetite for some part of their technology for which, until then, they had avoided. Third, they should expect to build bridges to the technical personnel. Later, when the technical people approach them about a technical dilemma that needs their business-risk guidance, they will have both a technical and a personal base on which to build.

As a final word about technical literacy, we want to remind you that every professional practice has a technical component in which general managers must be conversant. Not only are computer and systems design seminars in your future, so are accounting, purchasing, production, marketing, and human resource seminars. Many professional organizations realize the need to get the word out to the nontechnical managers of technical staffs. They have organized seminars or meetings specifically targeted to these peripheral but influential players. Therefore, there is no excuse for any manager to limit his technical literacy to his primary technical root.

12.3 CHANGE DYNAMICS

The central difficulty with improving your appreciation and skill managing change is that managing change is what a manager does. Not unlike profit awareness, competence with change dynamics is easily overlooked as "second nature" when in fact it should be seen and practiced as a learned skill.

The best way to make change management part of your repertoire is to reread Chapter 4, "Change Dynamics"; Chapter 5, "Useful Change Tools"; and Chapter 6, "Controlling Change," just before you start your next big, identifiable project. While you can apply these ideas to an ongoing activity, it is better to flex your new intellectual muscle mass on a fresh undertaking than on an existing, troubled one. That way, you will gain a better appreciation for change dynamics and what it adds to the mix of fundamental skills.

We have a little secret to tell you here. Because of cognitive limits (discussed in Chapter 8, "Decision Errors"), the authors themselves routinely refer to their and others' writings on change dynamics before we start new projects. While we might be marginally better at recalling bits and pieces of these concepts, whenever we face a new project we always reread the principal work in this area. Why? Because we recognize the power of documented procedures in maintaining the best practices. As any student of a complex intellectual practice (like theology) will tell you, there is a reason the pious study the Bible, the Torah, or the Koran their entire lives. They know that God expects they will forget much of it.

12.4 DOCUMENTATION BASICS

It is highly likely that the document systems you run into during your career will be pitifully incomplete and marginally functional. Most are. Your managerial authority, along with the surrounding organizational culture, will dictate what can be done to improve the situation. As a minimum, you should try to instill the discipline and organization outlined in Chapter 7 in the departments you personally control. The assembly of control elements into the artificial but useful "policy and procedure" dichotomy works at the five-person staff level as well as it does at the enterprise level.

More than likely, however, you will not come back to this fundamental skill until the documentation system you depend on fails. Why? Because of all the fundamental manageering skills, understanding the basics of documentation can be easily portrayed as a necessary evil instead of as an indispensable friend. Additionally, improving documentation is a change project in itself; therefore, you might not want to touch the subject until you can be sure that significant disconfirming information has accumulated such that the force field will favor change over the status quo.

But that is not all bad. Since poor documentation systems are so pervasive and at the root of many organizational efficiency problems, improving them has to be a companywide program—a fad, if you will. As we have said before, fads are not bad if they are put into place with enough understanding and built-in bias for change.

12.5 DECISION THEORY

Improving your facility with decision theory is probably the most enjoyable activity of the six. After all, it is a mind game. Similar to the idea we proposed in Section 12.3, "Change Dynamics," we would suggest that you also review Chapters 8 and 9 before you start your next big project. As you do so, you will find that certain techniques we outlined will apply to the task at hand. If you need more information about them before you try them out, check some of the references we list in the appendix. They will stimulate and encourage you to develop your decision techniques more than we could have in our simple introduction to the topic.

The interesting thing about studying decision theory is that many technically literate people find themselves applying what they learn more to their personal predicaments than to their business dilemmas. Such is the power of disciplined thinking.

12.6 DUALITY

Duality is another nontool-like fundamental skill similar to profit awareness. To help you get your arms around this feature of good management, we offered the inclusion tree as a simple but effective tool. While the inclusion tree is not duality in action, it forces you to consider the singularly critical element of duality, that of consistency. The tree also makes you slow down for curves. That is, unless something makes you slow down and consciously consider your duality, especially when you are most exposed, you will never do so until you are bitten by a poor duality outcome, at which time it is too late to recover. So whenever you organize a meeting, use the inclusion tree to help answer the question, "Whom do I invite, whom do I exclude?" Use it when you are considering asking someone to give you information you need to make a decision. "Whom do I ask, if anyone? Whom do I exclude? What will the information cost in terms of my duality?"

12.7 THE SUM IS GREATER THAN THE PARTS

Eventually, you will grow comfortable with all the fundamental skills of manageering. They will become part of the way you do business, but only a part. Every reader of this book will emphasize a chosen few over the others because he or she feels more comfortable with some "favorite" fundamentals. The chosen fundamentals fit that reader's individual personality better and therefore are more easily assimilated. But if everyone picked up these fundamentals in the same proportions, without modification, then the variety that is the spice of life (and the source of competitive advantage) would vanish. Obviously, varying degrees of emphasis are natural.

The consequences of this favoritism are not overwhelming for early career managers. But as managers mature and accept more responsibility for the overall health of their organizations, then their weaker fundamentals have to be addressed, or the enterprise will suffer. But like a hitter who cannot field or a catcher who cannot hit, you can build a strategy that overcomes your weaknesses so they do not diminish your effectiveness.

12.8 ONE MORE TRIP

Suppose you are hired as a division or company president to "clean up the store." Headquarters or the shareholders are not happy with the results of the last management team and they want things fixed. Of course, they do not really know what the problem is, even if they say they do. If they did, those problems would have been fixed by now, right?

Anyway, what is your strategy for applying the fundamentals of manageering? The place is a mess. They have been right-sized, downsized, empowered, reengineered, TQM'd. Nothing seems to have helped. Like the battlefield general who comes in to pick up the pieces after a colossal reversal of fortune, where do you start? Everywhere? Nowhere? Up there? Down here?

Suppose you decide to gather a few key people together and get some information. These people have been going to meetings until they are punch drunk. Half their peers have been fired. The last manager they had started out the same way, with a meeting. And you are going to call another meeting to gather more information? Luckily, you consulted the inclusion tree before you made any grand entrances. Not that you would not have held the meeting, but at least you are going to be sure that it will be more effective than anything they have seen before.

But in the end there is only one person who knows what to do: you. You know that your staff's fundamental business skills have deteriorated so badly that you have to build them up again. That is where maturity and experience in manageering comes in. Every staff, every division, every company is different. You most likely would not find the same combination of deteriorated skills in everyone; even if you did, the impact of that deterioration on effectiveness and profitability would differ from company to company. The fundamentals of manageering are a whole fabric, and that is why, to make it as a senior manager, your (personal) weakest fundamental can be no worse than that of your best middle manager.

That being said, deciding on where to start a rebuilding process is conditional on what you find on the ground. First, go back to decision theory. Remember about writing things down? You will make a list of the six fundamentals. Then you will spend a little unfocused time deciding which attributes you should measure to get a handle on which of your staff's fundamentals is the weakest. Then you will grade the unit and get a fix on the size and position of the problem.

Next, you will have to decide which fundamentals are critical to survival and need immediate attention and which can be safely put aside until you have time to deal with them. You will pencil in a program, but it will not be a lecture series. It will be actual application. You do not have time for lectures. You will teach by doing, which is another foundation of duality.

The point of this simple scenario is that while listing the fundamentals of manageering is a simple, straightforward exercise, applying them holistically in the real world is a complex activity that moves the actor from the world of prescription to the universe of integration. That is where your individuality and intellect, as opposed to background and training, begin to play a larger part in the development of leadership (as opposed to management) skills, and we, as experts in management development, have little intelligent left to say.

12.9 WHERE DO GOOD MANAGERS COME FROM?

Managers are born fully formed, not made. "No, no," some say. "You are wrong. Managers are made, not born." Which is it and why should you care?

You should care where managers come from. Since you have spent precious energy reading this book, you obviously have more than just a passing interest in your development as a general manager. At this point in your career, you are interested in how you might improve your management skills as well as those same skills in those people who follow you. If managers are more born than made, then efforts at management development should emphasize the art and science of identifying those so blessed. If managers are more made than born, the implication is that the pool of talent is large and those who become managers merely need the motivation to do so.

The truth, of course, lies somewhere in between and varies from individual to individual depending on the drive for success, interest in people and organizations, propensity for conflict, and ration of fate. As point in fact, it is not unusual for great leaders to have spates of mediocre performance. Eisenhower was an average West Point cadet. Grant was disappointed at his accomplishments in civilian life and in his undistinguished pre–Civil War Army career. Yet both were distinguished leaders.

For our purposes, the backgrounds of great leaders hold inspiration, but not lessons. At our level of endeavor, we are more interested in what separates the mediocre manager from the successful one, what separates you from the rest of the management tribe. In getting at that answer, we have to lay aside what we cannot control and focus on what we can. Ultimately, the question of "nurture versus nature" in the development of managers is irrelevant because it is unanswerable. However, if we have proved anything in this book, we hope it is that, all things being equal, management skills are learnable skills. The important variable in this education is the presentation of appropriate subject matter.

In business as in life, it is the fundamentals that count. We hope we have demonstrated those fundamentals well enough so you can apply them or, at the very least, so you have an appreciation that they really do count, regardless of whether your fundamentals are the ones we endorse or ones you develop elsewhere. Your grounding in your fundamental skill set will be your professional competitive advantage, since so many of your contemporaries will lose their way in the maze of popular management fads and quick fixes. We have seen many people build personal fortunes on the framework of a singular set of management fundamentals. You can do the same, whether in the end you remain in management or return to your technical-staff roots. Now all you need is a little luck.

Appendix

The reader is referred to the following texts for more information about the topics discussed in this book.

A.1 CHANGE DYNAMICS

Beckhard, Richard, and Reuben T. Harris, *Organizational Transitions: Managing Complex Change, 2nd Edition,* Reading, MA: Addison-Wesley, 1987.

A pamphlet-sized book that is surprisingly complete; for those who feel more comfortable cooking with a recipe.

Schein, Edgar H., "Planning and Managing Change," MIT Management in the 1990s Research Program Working Paper (#90s: 88-056). Boston: MIT Sloan School of Management, October 1988.

The authors' original source of ideas about the stages of the change process.

A.2 DECISION THEORY

Dawes, Robyn M., *Rational Choice in an Uncertain World,* San Diego: Harcourt Brace Jovanovich, 1988.

Heavy into the probabilistic nature of uncertain outcomes; an excellent treatment of the subject but requires concentrated effort to get the most from it.

Hogarth, Robin, *Judgement and Choice: The Psychology of Decision, 2nd Edition,* Chichester, Great Britain: John Wiley & Sons, 1987.

Overlaps Dawes somewhat but concentrates more on cognitive limitations. If you do not understand Dawes, try Hogarth or vice versa.

Janis, Irving L., *Groupthink: Psychological Studies of Policy Decisions and Fiascoes,* Boston: Houghton Mifflin, 1982.

A great read if you like inside stories of power, politics, and influential people screwing up because they are full of themselves.

A.3 DECISION-MAKING TOOLS

Behn, Robert D., and James W. Vaupel, *Quick Analysis for Busy Decision Makers,* New York: Basic Books, 1982.

Well-written how-to book about decision analysis, especially decision trees.

A.4 FADS

Covey, Stephen R., *The Seven Habits of Highly Effective People: Powerful Lessons in Personal Change,* New York: Simon & Schuster, 1989. [Also published with the subtitle *Restoring the Character Ethic.*]

Crosby, Philip B., *Quality Is Free: The Art of Making Quality Certain,* New York: McGraw-Hill, 1979.

Deming, Edwards W., *Out of the Crisis,* Cambridge, MA: MIT Center for Advanced Engineering Study, 1986. [Deming's style is a bit hard to read. You might try *The Deming Management Method,* by Mary Walton (New York: Putnam, 1986) for an easier read.]

Hammer, Michael, and James Champy, *Reengineering the Corporation: A Manifesto for Business Revolution,* New York: HarperCollins, 1993

Each of the four preceding books was the basis for the management fad it started. They are shown here because we refer to them elsewhere in the book; we do not endorse them.

A.5 GESTALT CYCLE OF EXPERIENCE

Nevis, Edwin C., *Organizational Consulting: A Gestalt Approach,* New York: Gardner Press, 1987.

Primarily written for process consultants, this book has a wealth of good ideas for general managers, including a detailed explanation of the Gestalt Cycle of Experience.

A.6 MEETINGS

Jay, Antony, "How To Run a Meeting," *Harvard Business Review,* March–April 1976.

Good advice is never dated. Still the best treatment of the subject.

A.7 QUALITY AS A FUNCTION

Feignenbaum, Armand V. *Total Quality Control, 3rd Edition,* New York: McGraw-Hill, 1991.

Juran, J. M., editor, *Juran's Quality Control Handbook, 4th Edition,* New York: McGraw-Hill, 1988.

These two books describe nearly all the elements of the practice of the quality control.

A.8 SELF-REFERENTIALISM

Guillen, Michael, *Bridges to Infinity: The Human Side of Mathematics,* Los Angeles: Jeremy P Tarcher, 1983.

A great bedtime read (really). Short, well-written chapters on profound mathematical and philosophical problems; accessible to anyone faintly curious about mathematics, physics, and the universe. Good treatment of Gödel's incompleteness theorem but no mention of quality.

About the Authors

Kenneth Durham graduated with honors from the University of Texas, where he earned a B.S. in electrical engineering. He also has an M.B.A. from the MIT Sloan School of Management. Mr. Durham was a cofounder of Quality International LLP and is a principal with the Quantum Institute. He has held chief operating officer and technical management positions in Fortune 500 companies as well as start-ups. His assignments have included work ranging from mineral extraction to high-alloy metalworking to low-tech manufacturing to pure service. He has authored a wide range of titles, including peer-reviewed engineering dissertations and general business topics.

Bruce Kennedy graduated from the University of Central Oklahoma with a B.A. in marketing. He is the president of the Quantum Institute, has held management positions in several Fortune 500 companies, and was a founding partner in three successful start-up companies. Mr. Kennedy is recognized as an authority in vendor relations, inventory management, logistics, and the use of standards-based management systems. He has addressed over 40 professional organizations and has held related seminars in over 60 leading corporations during the past three years. His writings about management issues have been published throughout the United States and Europe by numerous technical institutions, universities, trade organizations, and professional societies.

Index